USGBC LEED Green Associate Study Guide

USGBC LEED Green Associate Study Guide Acknowledgments:
The USGBC LEED Green Associate Study Guide is a valuable tool for exam candidates planning to attain the Green Associate credential. We would like to extend our deepest gratitude to those involved in the production of this resource.

PROJECT TEAM

Green Building Services, Inc. (GBS)
Alicia Snyder-Carlson, *Content Developer*
Beth Shuck, *Content Developer*
Caitlin Francis, *Project Manager and Content Developer*
Glen Phillips, *Technical Specialist and Content Developer*
Katrina Shum Miller, *Principal in Charge*

Institute for the Built Environment (IBE)
Brian Dunbar, *Educational Consultant*
Josie Plaut, *Educational Consultant*

NonObvious Solutions
Elizabeth Gast, *Technical Illustrator and Graphic Designer*
Eric von Schrader, *Instructional Designer*

Prometric
Examination Question Writing Training

LEED Curriculum Committee Members
Draft reviews

USGBC Staff
Karol Kaiser, *Director of Education Development*
Jacob Robinson, *Project Manager*

CONTENTS

LEED GREEN ASSOCIATE

This study guide is a resource to help you prepare for the LEED Green Associate Examination. It summarizes the critical points of green design, construction, and operations. To help you master its content, the guide has been packaged with the Green Building and LEED Core Concepts Guide, one of the reference documents for the LEED Green Associate Exam. Within each category of the guide, you will find a variety of study tools, including category reviews, review questions and worksheets, learning activities, and practice questions.

If you have decided that the LEED Green Associate credential is right for you, then you may already understand that by pursuing this credential you will be exposed to a world of opportunities.

Your LEED Green Associate credential will help you stay competitive as organizations seek more knowledgeable employees for workforces grounded in sustainability principles that can be applied across business disciplines.

Congratulations on your decision to pursue the LEED Green Associate credential. Best of luck on the exam!

Getting Started on Your LEED Green Associate Credential

Earning the LEED Green Associate Credential requires passing a two-hour exam composed of 100 questions. This exam attests to your knowledge of good environmental practice and skill and reflects your understanding and support of green design, construction, and operations.

STEP 1: Read the GBCI LEED Green Associate Candidate Handbook.

(Free for download at www.gbci.org.) Determine whether you meet the eligibility requirements. To take the LEED Green Associate Exam, you must meet one of the following:

- Have experience in the form of involvement on a LEED-registered project;
- Be employed (or have been previously employed) in a sustainable field of work; or
- Be enrolled in (or have completed) an education program that addresses green building principles.

STEP 2: Register for and schedule your exam.

STEP 3: Access the reference documents.

The Candidate Handbook lists the primary and ancillary references that are the sources for the exam questions. The "Green Building and LEED Core Concepts Guide" is available for purchase through USGBC, and the remaining references are hyperlinked and available for free download through the LEED Green Associate Candidate Handbook on the GBCI website. Exam reference documents are subject to change as the GBCI exams evolve. Always check the candidate handbook for the most up-to-date list.

Primary References*

- *Green Building & LEED Core Concepts Guide, 1st Edition* (US Green Building Council, 2009) (available for purchase at www.usgbc.org/store)

- *Green Office Guide: Integrating LEED Into Your Leasing Process, Section 2.4* (US Green Building Council, 2009)

- *LEED 2009 for New Construction and Major Renovations Rating System* (U.S. Green Building Council, 2009)

- *LEED for Existing Buildings: Operations & Maintenance Reference Guide*, Introduction (U.S. Green Building Council, 2008)

- *LEED for Existing Buildings: Operations & Maintenance Reference Guide*, Glossary (USGBC, 2008)

- *LEED for Homes Rating System* (USGBC, 2008)

- *Cost of Green Revisited*, by Davis Langdon (2007)

- *Sustainable Building Technical Manual: Part II*, by Anthony Bernheim and William Reed (1996)

- *The Treatment by LEED® of the Environmental Impact of HVAC Refrigerants* (LEED Technical and Scientific Advisory Committee, 2004)

- *Guidance on Innovation & Design (ID) Credits* (USGBC, 2004)

- *Guidelines for CIR Customers* (USGBC, 2007)

*Note: You should also be familiar with the content of the U.S. Green Building Council's Website, www.usgbc.org, including but not limited to LEED Project Registration, LEED Certification content, and the purpose of LEED Online. The U.S. Green Building Council's LEED Website, www.usgbc.org/leed, also has free access to LEED Rating Systems, LEED Reference guide Introductions, and Checklists beyond those listed above.

You will also find a list of abbreviations and acronyms in the LEED for Homes Rating System on pages 105–106 and a helpful glossary of terms on pages 107–114.

Ancillary References

* *Energy Performance of LEED® for New Construction Buildings: Final Report*, by Cathy Turner and Mark Frankel (2008)

* *Foundations of the Leadership in Energy and Environmental Design Environmental Rating System: A Tool for Market Transformation* (LEED Steering Committee, 2006)

* *AIA Integrated Project Delivery: A Guide* (www.aia.org)

* *Review of ANSI/ASHRAE Standard 62.1-2004: Ventilation for Acceptable Indoor Air Quality*, by Brian Kareis (www.workplacegroup.net)

* *Best Practices of ISO - 14021: Self-Declared Environmental Claims*, by Kun-Mo Lee and Haruo Uehara (2003)

* *Bureau of Labor Statistics* (www.bls.gov)

* *International Code Council* (www.iccsafe.org)

* *Americans with Disabilities Act (ADA): Standards for Accessible Design* (www.ada.gov)

* *GSA 2003 Facilities Standards* (General Services Administration, 2003)

* *Guide to Purchasing Green Power* (Environmental Protection Agency, 2004)

* *LEED 2009 for Operations & Maintenance Rating System* (USGBC, 2009)

STEP 4: Start studying!

The Green Building and LEED Core Concepts Guide includes many of the key concepts from the primary and ancillary reference documents noted above, but you should still access all the reference documents and review them to be adequately prepared for the exam.

 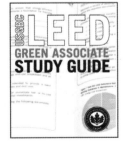 

LEED GREEN ASSOCIATE

I. Content Areas

The Green Associate Exam has seven major areas of focus, which are called out in the Candidate Handbook. Here is how they align with the LEED rating system credit categories:

GBCI EXAM AREAS OF FOCUS		LEED RATING SYSTEM CREDIT CATEGORIES
I.	**Project Site Factors** =	Sustainable Sites (SS)
II.	**Water Management** =	Water Efficiency (WE)
III.	**Project Systems and Energy Impacts** =	Energy and Atmosphere (EA)
IV.	**Acquisition, Installation, and Management of Project Materials** =	Materials and Resources (MR)
V.	**Improvements to the Indoor Environment** =	Indoor Environmental Quality (IEQ)
VI.	**Stakeholder Involvement in Innovation** =	**Innovation in Design (ID)** & **Regional Priority (RP)**
VII.	**Project Surroundings and Public Outreach** =	

II. Exam Questions

The Green Associate Exam questions are:

- Developed and validated by global work groups of subject matter experts;

- Referenced to current standards and resources;

- Developed and monitored through psychometric analysis; and

- Designed to satisfy the test development specifications of a job analysis.

The questions assess your knowledge at three levels:

- **Recall questions** test your direct knowledge of concepts. This section may require you to define terms or concepts, recall facts, recognize or identify components or steps in a process, and group items into categories.

- **Application questions** evaluate your knowledge of procedures and performance and may require you to demonstrate how things work, perform calculations following a formula, place components or steps into proper sequence, describe how a process works, and apply a known process or sequence of actions to accomplish a task (such as troubleshooting a problem using a detailed checklist).

- **Analysis questions** test your reasoning and problem-solving abilities. Such questions may require you to demonstrate an understanding of how things work, cause and effect, and underlying rationale; analyze problems and devise appropriate solutions; build a conceptual model of a process; troubleshoot a problem without a checklist; and analyze and solve a problem.

The exam questions follow consistent formats:

- You will never see an "all of the above," "none of the above," or "true/false" type of question on this test, because:
  - These questions can cause confusion and have overlapping answers.
  - The test is intended to be clear and straightforward.
  - The question language is never intended to be tricky.

- You will never see a credit number listed by itself. Any direct reference to a LEED credit will include the full credit name.

- Most acronyms are spelled out so that you do not need to memorize all acronyms you learn.
  - Commonly referenced acronyms may be used (i.e. LEED, ASHRAE, and VOC) so it is still a good idea to know what these acronyms stand for!

PRACTICE QUESTIONS

Practice questions in this guide were written by subject-matter experts trained by Prometric, the testing company that administers the GBCI LEED exams, to use the same guidelines as the item writers for the actual examinations. These practice questions will allow you to become familiar with the exam expectations, format, and question type. This approach should improve your testing skills and alleviate stress on test day, allowing you to focus on core information.

STUDY TIPS

You will learn best if you establish a regular study schedule over a period of time. Daily studying in shorter sessions is more effective for most people than "cramming" in long sessions at the last minute.

Studying with a partner or a group can help you stay on schedule and give you opportunities to quiz and drill with each other. Group learning activities are provided throughout this guide to give you ideas for how to study as a group.

Here's a step-by-step approach for using your study resources:

- Read the Green Building and LEED Core Concepts Guide one category at a time. Don't try to learn everything on the first pass.

- Read the corresponding section in this study guide appendix.

- Take notes and highlight key points.

- Review the other reference materials that apply to the category.

- Reread the category.

- Utilize the review questions, learning activities, and practice questions in this guide.

- Keep reviewing and rereading until you are confident you know the material.

EXAM DAY TIPS

General Strategies

- Always arrive early and take a moment to relax and reduce your anxiety.
    - This brief time period will boost your confidence.
    - Use this time to focus your mind and think positive thoughts.

- Plan how you will use the allotted time.
    - Estimate how many minutes you will need to finish each test section.
    - Determine a pace that will ensure that you complete the whole test on time.
    - Don't spend too much time on each question.

- Maintain a positive attitude.
    - Don't let more difficult questions raise your anxiety and use your valuable time. Move on and find success with other questions.
    - Avoid watching for patterns. Noticing that the last four answers are "c" is not a good reason to stop, go back, and break concentration.

- Rely on your first impressions.
    - The answer that comes to mind first is often correct.
    - Nervously reviewing questions and changing answers can do more harm than good.

- Plan to finish early and have time for review.
    - Return to difficult questions you marked for review.
    - Make sure you answered all questions.

Multiple Choice Strategies

- Formulate your own answer before reading the options.
    - Cover up the answer options and see if you can answer the question without looking at the options. Focus on finding an answer without the help of the alternative options.

- This process will increase your concentration.

- Doing this will help you exercise your memory.

- Read all the choices before choosing your answer.

- Eliminate unlikely answers first.
  - Eliminating two alternatives quickly may increase your probability to 50-50 or better.
- Look for any factor that will make a statement false.
  - It is easy for the examiner to add a false part to an otherwise true statement.
  - Test takers often read an answer and see some truth and quickly assume that the entire statement is true. For example, "Water boils at 212 degrees in Denver." Water does boil at 212 degrees, but not at Denver's altitude.
- Beware that similar answers provide a clue. One of them is correct; the other is disguised.
  - This is not a trick, but make sure you know the exact content being asked.
- Consider the answers carefully. If more than one answer seems correct for a single-answer question:
  - Ask yourself whether the answer you're considering completely addresses the question.
  - If the answer is only partly true or is true only under certain narrow conditions, it's probably not the right answer.
  - If you have to make a significant assumption in order for the answer to be true, ask yourself whether this assumption is obvious enough that everyone would make it. If not, ignore that answer.
- If you suspect that a question is a trick item, make sure you're not reading too much into the question, and try to avoid imagining detailed scenarios in which the answer could be true. In most cases, "trick questions" are only tricky because they're not taken at face value.
  - The test questions will include only relevant content and are not intended to trick you or test your reading ability.

WHAT ABOUT GREEN BUILDING?

WHY BUILD GREEN?

- What percentage of electricity consumption do buildings account for in the United States?

- Will the value of green building continue to increase in the coming years?

- What is the value of assessing the life of a building and investing in the long-term performance of that space?

- Is green building about new technologies or getting back to the basics of good design? Or both? How can these elements complement each other?

Read Pages 1–16 of the Green Building and LEED Core Concepts Guide, Second Edition, or pages 1-13 in the First Edition.

GREEN BUILDING OVERVIEW

In this chapter, you will learn about how green building transforms the way buildings are designed, built, and operated to:

- Create more comfortable, healthier, and sustainable built environments;
- Create these built environments while reducing energy consumption, greenhouse gas emissions, water consumption, and solid waste generation; and
- Reduce costs, reduce liability, increase value, and achieve more predictable results in the design, construction, and operation of built environments.

The cumulative impact of the design, construction, and operation of built environments has profound implications for human health, the environment, and the economy. Three key issues were discussed in this section:

- Life cycle of built environments;
- Integrative approach; and
- Green building costs and benefits.

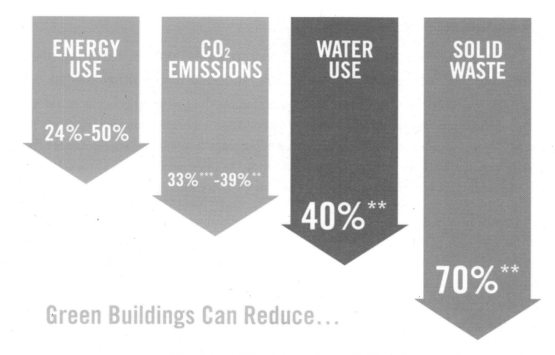

ENERGY USE
24%-50%

CO₂ EMISSIONS
33%***-39%**

WATER USE
40%**

SOLID WASTE
70%**

Green Buildings Can Reduce...

* Turner, C. & Frankel, M. (2008). Energy performance of LEED for New Construction buildings: Final report.
** Kats, G. (2003). The Costs and Financial Benefits of Green Building: A Report to California's Sustainable Building Task Force.
*** GSA Public Buildings Service (2008). Assessing green building performance: A post occupancy evaluation of 12 GSA buildings.

LIFE CYCLE OF BUILT ENVIRONMENTS

This section discusses both the life cycle of the building itself and the life cycle of the cost of the building.

When considering the life cycle of a building, think about the building in a holistic sense. Consider things like where building materials come from and where they will go after use, what the energy and water sources are and how those affect the building's surroundings, and how the building users will get to and from the building.

This concept takes green building to the next level and encourages project teams to create environments that truly regenerate their surroundings, providing positive environmental, social, and economic impacts.

Life-cycle cost analysis looks at the cost of green building and what types of advantages are discovered when initial investments are made in more durable products and efficient building systems.

INTEGRATIVE APPROACH

In conventional design and construction processes, project team members move through each phase relatively independently of the other design disciplines, with little communication. For example, the civil designer works on a way to ensure that all stormwater is directed away from the building to a storm system, while the landscape architect works on an irrigation design for all the new plantings. If those two disciplines were better coordinated, the landscape architect might be able to tie the irrigation system into a rainwater collection system, greatly reducing the amount of potable water needed.

Think about areas of opportunity for integration and synergy throughout the design and construction process. Keep in mind that many design decisions associated with environmental impacts are made in the first part of the design process. Therefore, input from all key stakeholders and members of the design team is essential before schematic design begins.

```
                                                              Substantial Completion
PREDESIGN  DESIGN                      BID   CONSTRUCTION           OCCUPANCY
              SD |DD |CD                      CA
```

Construction Phases

- **Predesign** entails gathering information, recognizing stakeholder needs, establishing project goals, and selecting the site.

- **Design**
  - **Schematic design (SD)** explores several design options and alternatives, with the intent to establish an agreed-upon project layout and scope of work.
  - **Design development (DD)** begins the process of spatial refinement and usually involves the first design of a project's energy systems.
  - **Construction documents (CD)** carry the design into the detail level for all spaces, systems, and materials so that construction can take place.

- **Bidding** is when costs are established and contracts for construction services are signed.

- **Construction, or construction administration (CA),** involves the actual construction of the project. Commissioning takes place near the end of construction, once the systems have been installed and are operable.
  - **Substantial completion** is a contractual benchmark that usually corresponds to the point at which a client could occupy a nearly completed space.
  - **Final completion**.
  - **Certificate of occupancy** is the official recognition by a local building department that a building conforms to applicable building and safety codes.

- **Occupancy** occurs after the certificate of occupancy has been received and is continued throughout the life of the building. During occupancy, periodic maintenance must occur. Additionally, recommissioning along with occupant surveying (via a post-occupancy evaluation) should take place at regular intervals.

Over the duration of the design and construction process, several parties are involved, making decisions and affecting the project in different ways. Think about the responsibility each of these roles has and how their partnership, communication, and integrated approach may play a role in the success of the project.

Players Involved

- **The project owner** defines the intended function of the building and selects the primary members of the project team.

- **The architect** holds the primary responsibility for the design of the building as a whole and is responsible for coordinating all of the design team.

- **Engineers** are responsible for the design of building systems and other technical elements, such as mechanical systems, electrical systems, and plumbing.

- **The commissioning authority** leads and oversees the commissioning process.

- **The general contractor** is responsible for construction of the building and the work of the subcontractors.

- **Facilities staff** maintain the building once it is complete and are often engaged during the design and construction process to ensure that their operational needs are met.

- **Building users** are occupants and users of the completed building. Meeting their needs should be a primary focus of the design, construction, and operations efforts.

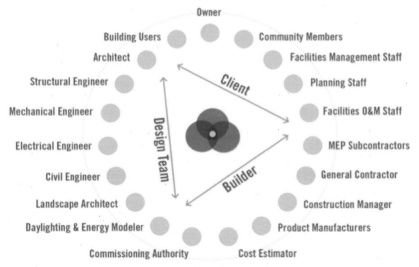

Adapted from graphic by Bill Reed

Integrative Approach: Key Stakeholders. The concept of integrative design emphasizes connections and communication between professionals and throughout the life of a project. Input from all key stakeholders and members of the design team is essential before schematic design begins, particularly since 70% of the decisions associated with environmental impacts are made during the first 10% of the design process. Not all of the team members indicated in this diagram participate in all LEED projects, but each is responsible for a system or components that impact nearly all others.

In a conventional design process, the systems may be chosen and installed independently. The outcome is that each piece may be working according to its specifications, but, put together, the pieces don't work as optimally as they could.

CONVENTIONAL BUILDING PROCESS

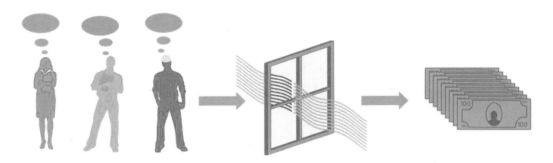

Independent design of:
- HVAC
- Insulation
- Window systems

- Large HVAC system
- Minimal insulation
- Less effective insulation

Higher first cost and operating costs

VS.

INTEGRATIVE DESIGN PROCESS

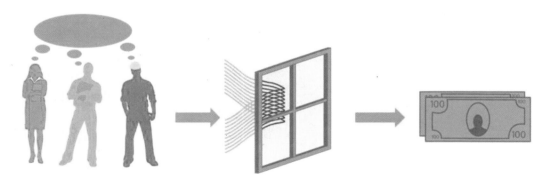

Integrative design of:
- HVAC
- Insulation

- Smaller HVAC system
- More effective insulation
- High performance windows

Lower first cost and operating costs

GREEN BUILDING COSTS AND BENEFITS

Building green doesn't cost more! Studies demonstrate an actual average marginal cost of less than 2% associated with green building. As "The Cost of Green Revisited," a 2007 study by construction consulting firm Davis Langdon, points out, many project teams are building green buildings with little or no added cost, and with budgets well within the cost range of non-green buildings with similar programs. The study also found that, in many areas of the country, the contracting community has embraced sustainable design and no longer sees sustainable-design requirements as additional burdens to be priced in their bids.

Be sure to check out the websites listed in the primary references for more information on the cost of building green.

GREEN BUILDING CATEGORY REVIEW

What factors motivate green building?

What are the differences between conventional design and integrative design?

How can green buildings address environmental issues?

What are the added costs of building green, and what benefits offset these costs?

LEARNING ACTIVITIES

WRITE IT DOWN
List the benefits of green building that are important to you. Write down every benefit you can think of; then, try to put them in order from most important to least important. How does this exercise make you think differently about green building?

THINK ABOUT IT
Think of three buildings with which you are familiar (such as your home, your place of work, a school, or a store you visit frequently). Write down your ideas on these questions:

- What about them makes them green? What makes them not so green?

- In what ways could each of these buildings become greener?

- What costs or other challenges would have to be addressed to make them greener?

INVESTIGATE
Find the "greenwash." Locate two advertisements that make claims relative to the sustainable attributes of the product or service being advertised. Review each ad and identify which claims are legitimate and which are illegitimate or unsubstantiated.

DISCUSS IN A GROUP
Find an image of a green building. Take five minutes to look over the image and independently identify or visualize as many green building strategies as you can. Share these as a group. Then, work to identify which category the strategies fit best in: environmental, economic, or social. Identify those that overlap.

PRACTICE QUESTIONS

1. **What constitutes the largest use of energy in buildings in the United States?**

   a.) Space cooling
   b.) Space heating
   c.) Electric lighting
   d.) Water heating

2. **Green building emphasizes using what type of design process?**

   a.) Linear
   b.) Multistage
   c.) Integrative
   d.) Tiered

3. **What are the three dimensions of sustainability often described as the triple bottom line or the three-legged stool?**

   a.) Economic prosperity, environmental stewardship, and social responsibility
   b.) Economic theory, cultural agendas, and global variety
   c.) Energy efficiency, water efficiency, and indoor environmental quality
   d.) Government standards, building codes, and building practices

4. **Life-cycle assessment is used to determine the ___?**

   a.) Balance of natural cycles such as the hydrologic cycle
   b.) Environmental aspects and potential impacts of a given product
   c.) Life span of a building and its components
   d.) Environmental systems affected over the life of a building

5. **When is the best time to incorporate an integrative approach for a building project?**

   a.) Predesign
   b.) Schematic design
   c.) Design development
   d.) Construction documents
   e.) Construction

6. **Credit weightings are based on ___?**

   a.) Relative costs and benefits of each credit
   b.) Environmental impacts and human benefits
   c.) Carbon footprint and embodied energy
   d.) Expected environmental performance

7. **Implementation of green building strategies such as daylighting, passive cooling, high-efficiency mechanical systems, and stack ventilation contributes to what type of cost savings?**

   a.) Reduced first costs
   b.) Reduced maintenance costs
   c.) Reduced life-cycle costs
   d.) Reduced end-of-life costs
   e.) Reduced imminent costs

8. **The installation of low-flow faucet aerators at hand washing stations will result in which of the following (select two)?**

   a.) Reduced hours of occupancy
   b.) Reduced energy use
   c.) Reduced water use
   d.) Increased occupant productivity
   e.) Reduced stormwater runoff

Answer Key on Page 133

KEY TERMS TO KNOW:

Cover up the left side of the page and test yourself to see if you can summarize the definitions of the following key terms. Or better yet, make flash cards.

COPY   CUT   KEY TERM   DEFINITION   FOLD   KEY TERM   REVIEW

Key Term	Definition
Biodegradable	Capable of decomposing under natural conditions. (EPA)
Carbon Footprint	A measure of greenhouse gas emissions associated with an activity. A comprehensive carbon footprint includes building construction, operation, energy use, building-related transportation, and the embodied energy of water, solid waste, and construction materials.
Ecosystem	A basic unit of nature that includes a community of organisms and their nonliving environment linked by biological, chemical and physical processes.
Environmental Sustainability	Long-term maintenance of ecosystem components and functions for future generations. (EPA)
High-Performance Green Building	A structure designed to conserve water and energy; use space, materials, and resources efficiently; minimize construction waste; and create a healthful indoor environment.

Integrated Design Team	All the individuals involved in a building project from early in the design process, including the design professionals, the owner's representatives, and the general contractor and subcontractors.
Life-Cycle Assessment	An analysis of the environmental aspects and potential impacts associated with a product, process, or service.
Market Transformation	Systematic improvements in the performance of a market or market segment. For example, EPA's ENERGY STAR program has shifted the performance of homes, buildings, and appliances toward higher levels of energy efficiency by providing recognition and comparative performance information through its ENERGY STAR labels.
Regenerative Design	Sustainable plans for built environments that improve existing conditions. Regenerative design goes beyond reducing impacts to create positive change in the local and global environments.
Sustainability	Meeting the needs of the present without compromising the ability of future generations to meet their own needs. (Brundtland Commission)

WHAT ABOUT THE U.S. GREEN BUILDING COUNCIL AND ITS PROGRAMS?

- Can a building perform better than standard? If there are local and national building codes, why do we need another building standard?

- Are architects, engineers, and contractors trained in green building principles, either in school or on the job?

Read Pages 87–98 of the Green Building and LEED Core Concepts Guide, Second Edition, or pages 15-24 in the First Edition.

USGBC AND ITS PROGRAMS IN REVIEW

In this chapter, you will learn about the U.S. Green Building Council and its mission; Leadership in Energy and Environmental Design (LEED); the various LEED rating systems and their structure; as well as the Green Building Certification Institute (GBCI); the certification process; and credit interpretations. It is important to understand how USGBC and GBCI operate and the responsibilities they both have.

GREEN BUILDING®
CERTIFICATION INSTITUTE

	USGBC	**GBCI**
Organizational Overview	The U.S. Green Building Council (USGBC) is a 501(c)(3) nonprofit entity composed of leaders from every sector of the building industry working to promote buildings and communities that are environmentally responsible, profitable, and healthy places to live and work.	The Green Building Certification Institute (GBCI), established in January 2008, provides third-party project certification and professional credentials recognizing excellence in green building performance and practice.
Mission	"To transform the way buildings and communities are designed, built, and operated, enabling an environmentally and socially responsible, healthy, and prosperous environment that improves the quality of life."	"To support a high level of competence in building methods for environmental efficiency through the development and administration of a formal program of certification and recertification."
Primary Functions	<ul><li>Developed the LEED (Leadership in Energy and Environmental Design) Green Building Rating System. The LEED Green Building Rating System is the nationally accepted benchmark for the design, construction, and operation of high-performance green buildings.</li><li>Provides and develops LEED-based education and research programs.</li></ul>	<ul><li>Provides third-party LEED project certification.</li><li>Provides third-party LEED professional credentials.</li></ul>

Review of the Organization of the LEED Rating Systems

To successfully work on a LEED project, it is critical to understand how the LEED rating systems are structured and organized. Remember that prerequisites are required and credits are optional strategies that project teams pursue to achieve certification.

- **Prerequisites:** All must be met before a project can be certified, but they don't count toward point totals.
- **Credits are optional; meeting them adds points toward certification.**

For each certification level, a team must achieve a certain amount of credits in addition to all prerequisites.

The prerequisites and credits from all the LEED rating systems have been aligned, drawing on their most effective aspects. They are consistent across all LEED rating systems.

Each rating system comprises the following categories, except where noted. The prerequisites and credits fall within each of these categories:

 Sustainable Sites (SS)

 Location and Linkages (LL)
LEED for Homes Rating System only

 Water Efficiency (WE)

 Awareness and Education (AE)
LEED for Homes Rating System only

 Energy and Atmosphere (EA)

 Smart Location and Linkage (SLL)
LEED for Neighborhood Development Rating System only

 Materials and Resources (MR)

 Neighborhood Pattern and Design (NPD)
LEED for Neighborhood Development Rating System only

 Indoor Environmental Quality (IEQ)

 Green Infrastructure and Buildings (GIB)
LEED for Neighborhood Development Rating System only

 Innovation in Design (ID)/ Innovation in Operations (IO)

Anatomy of a Prerequisite and Credit

Each prerequisite and credit has an "intent," which identifies the prerequisite's or credit's main sustainability goal or benefit. Each prerequisite and credit also has at least one "requirement," which specifies the criteria that satisfy the prerequisite or credit and the number of points available. (Please see the sample credit in the appendix of this guide.)

Each prerequisite and credit in the LEED 2009 reference guides contains the following components:

- Credit name and point value;
- Intent;
- Requirements;
- Benefits and issues to consider;
- Related credits;
- Referenced standards;
- Implementation;
- Time line and team;
- Calculations;
- Documentation guidelines;
- Examples;
- Exemplary performance;
- Regional variations;
- Operations and maintenance considerations;
- Resources; and
- Definitions.

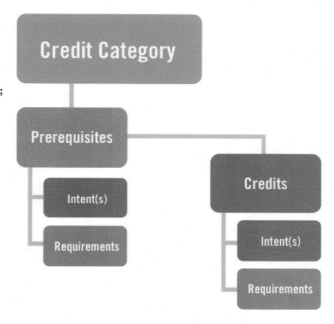

Certification Levels

Typically, each LEED rating system has 100 base points; Innovation in Design (or Operations) and Regional Priority credits provide opportunities for up to 10 bonus points. There are exceptions to the total amount of points possible; for example, LEED for Homes offers a total of 125 points.

- All LEED credits are worth a minimum of one point.
- All LEED credits are positive, whole numbers; there are no fractions or negative values.
- All LEED credits receive a single, static weight in each rating system; there are no individualized scorecards based on project location.

Refer to the appendix of this guide for a sample checklist for each of the LEED rating systems.

Certified	Silver	Gold	Platinum
40-49 points	50-59 points	60-79 points	80+ points

LEED Credit Weightings

The weighting and level of importance of a LEED credit depends on the credit's ability to address various environmental and human health concerns, such as smog formation and ozone depletion. Each credit was assigned its value based on how well the credit addresses these concerns, with the highest value given to those credits that offer the greatest potential benefit. LEED awards the most points for strategies that will benefit climate change and indoor environmental quality, especially those strategies that promote increased energy efficiency and reduced carbon dioxide (CO_2) emissions.

In the graphic below, the left column illustrates the issues LEED addresses, while the right column depicts their relative weighting in the LEED rating systems.

IMPACT CATEGORIES

CLIMATE CHANGE
INDOOR ENVIRONMENTAL QUALITY
RESOURCE DEPLETION
HUMAN HEALTH CRITERIA
WATER INTAKE
HUMAN HEALTH-CANCEROUS
ECOTOXICITY
EUTROPHICATION
HABITAT ALTERATION
HUMAN HEALTH-NONCANCEROUS
SMOG FORMATION
OZONE DEPLETION
ACIDIFICATION

→

CLIMATE CHANGE
INDOOR ENVIRONMENTAL QUALITY
RESOURCE DEPLETION
HUMAN HEALTH CRITERIA
WATER INTAKE
HUMAN HEALTH-CANCEROUS
ECOTOXICITY
EUTROPHICATION
HABITAT ALTERATION
HUMAN HEALTH-NONCANCEROUS
SMOG FORMATION
OZONE DEPLETION
ACIDIFICATION

The LEED Rating Systems

LEED addresses the different project development and delivery processes that exist in the U.S. building design and construction market through rating systems for specific building typologies, sectors, and project scopes:

- LEED for New Construction and Major Renovations
- LEED for Existing Buildings: Operations & Maintenance
- LEED for Commercial Interiors
- LEED for Core & Shell
- LEED for Homes
- LEED for Schools
- LEED for Retail
- LEED for Healthcare
- LEED for Neighborhood Development

Reference guides have been developed to aid in the implementation and understanding of the specific rating systems. The chart below shows how the rating systems are covered in each of the reference guides.

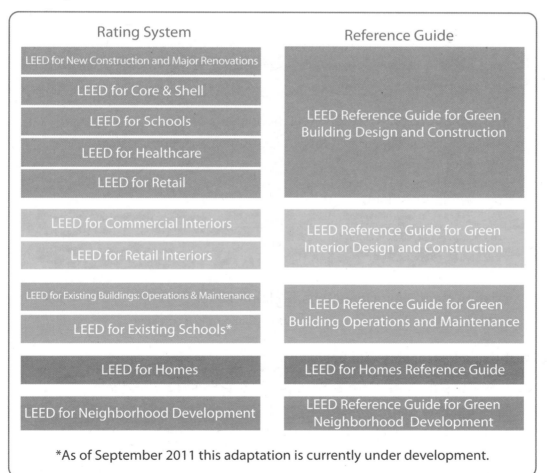

Rating System	Reference Guide
LEED for New Construction and Major Renovations	LEED Reference Guide for Green Building Design and Construction
LEED for Core & Shell	
LEED for Schools	
LEED for Healthcare	
LEED for Retail	
LEED for Commercial Interiors	LEED Reference Guide for Green Interior Design and Construction
LEED for Retail Interiors	
LEED for Existing Buildings: Operations & Maintenance	LEED Reference Guide for Green Building Operations and Maintenance
LEED for Existing Schools*	
LEED for Homes	LEED for Homes Reference Guide
LEED for Neighborhood Development	LEED Reference Guide for Green Neighborhood Development

*As of September 2011 this adaptation is currently under development.

What Project Types Are Eligible for LEED?

A project must adhere to the LEED Minimum Program Requirements (MPRs). These are critical! If non-compliance is discovered at any point, a project risks losing its certification and all associated fees that were paid!

Must Comply With Environmental Laws:

New Construction, Core & Shell, Schools, Commercial Interiors:
Do you know the environmental laws in your area? You must comply with all applicable federal, state, and local environmental laws and regulations in place where the project is located and at the time of design and construction.

Existing Buildings: Operations & Maintenance:
Not only the project, but also all normal building operations, must comply with all applicable federal, state, and local building-related environmental laws and regulations in place where the project is located from performance period through expiration date of the LEED Certification.

Must Be A Complete, Permanent Building or Space:

ALL Rating Systems:
You must design for, construct and operate on, a permanent location on already existing land. No moving allowed! If a building is designated to move at any point in its lifetime they cannot pursue LEED certification.

New Construction, Core & Shell, Schools:
You must include new, ground-up design & construction or major renovation of at least one building in its entirety.

Commercial Interiors:
The LEED project scope has to be distinct from all other spaces within the building with regards to at least one of the following: ownership, management, lease, or party wall separation.

Existing Buildings: Operations & Maintenance:
LEED projects must include at least one existing building in its entirety.

Must Use a Reasonable Site Boundary:

New Construction, Core & Shell, Schools, Existing Buildings: Operations & Maintenance:
Reasonable site boundary means that the LEED project must include all contiguous land. If it is on a campus, it must have project boundaries such that if all buildings on campus

become LEED certified then 100% of the gross land area on the campus would be included in the LEED boundary. The project boundary must not include land that is owned by a party other than that which owns the LEED projects, it must not unreasonably exclude sections of land to create boundaries in unreasonable shapes for the sole purpose of complying with prerequisites or credits, and any parcel of property may only be attributed to a single building.

Commercial Interiors:

You must include any land disturbed for the purpose of undertaking the LEED project in the LEED project boundary.

Must Comply With Minimum Floor Area Requirements:

New Construction, Core & Shell, Schools, Existing Buildings: Operations & Maintenance:

You must have at lease 1,000 square feet (93 square meters) of gross floor area.

Commercial Interiors:

You must have at lease 250 square feet (22 square meters) of gross floor area.

Must Comply With Minimum Occupancy Rates:

New Construction, Core & Shell, Schools, Commercial Interiors:

You must serve one or more Full Time Equivalent (FTE) occupants.

Existing Buildings: Operations & Maintenance:

Must be in a state of typical physical occupancy, and all building systems must be operating at a capacity necessary to serve the current occupants.

Must Commit to Sharing Whole Building Energy and Water Usage Data:

ALL Rating Systems:

You must commit to sharing with USGBC and/or GBCI all available actual whole project energy and water usage data for a period of at least 5 years. This commitment will carry forward if the building changes ownership and commences when occupancy occurs.

Must Comply With Minimum Building Area to Site Area Ratio:

ALL Rating Systems:

Gross floor areas must be no less than 2% of the gross land area within the LEED project boundary.

When to Use Each Rating System

As you learn about the application of each LEED rating system, keep in mind that some projects clearly fit the defined scope of only one LEED rating system; others may be eligible for two or more. The project is a viable candidate for LEED certification if it can meet all prerequisites and achieve the minimum points required in a given rating system. If more than one rating system applies, the project team can decide which to pursue.

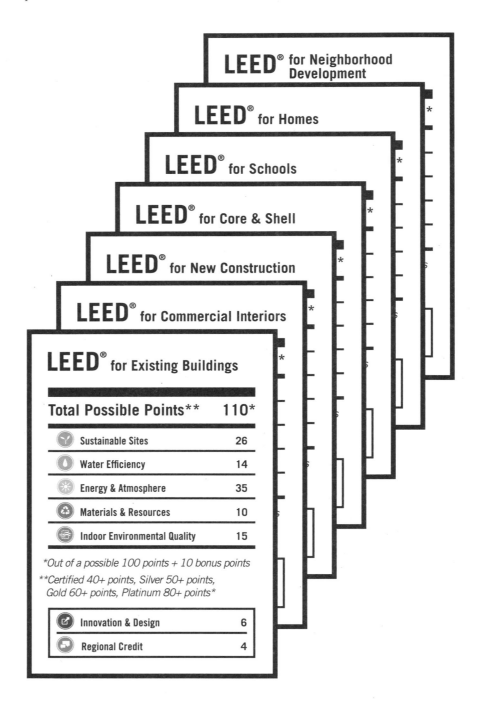

LEED® for Neighborhood Development

LEED® for Homes

LEED® for Schools

LEED® for Core & Shell

LEED® for New Construction

LEED® for Commercial Interiors

LEED® for Existing Buildings

Total Possible Points**	110*
Sustainable Sites	26
Water Efficiency	14
Energy & Atmosphere	35
Materials & Resources	10
Indoor Environmental Quality	15

*Out of a possible 100 points + 10 bonus points
**Certified 40+ points, Silver 50+ points,
Gold 60+ points, Platinum 80+ points*

Innovation & Design	6
Regional Credit	4

When to Use LEED for New Construction and Major Renovations

LEED for New Construction and Major Renovation was designed primarily for new commercial office buildings, but it has been applied to many other building types, as well.

- All commercial buildings, including offices, institutional buildings (libraries, museums, churches, and the like), and hotels, as well as residential buildings of four or more habitable stories are eligible.

- Design and construction activities for both new buildings and major renovations of existing buildings. A major renovation involves major HVAC renovation, significant envelope modifications, and major interior rehabilitation. If the project scope does not involve significant design and construction activities and focuses more on operations and maintenance activities, LEED for Existing Buildings: Operations & Maintenance is more appropriate.

- The owner or tenant must occupy more than 50% of the building's leasable square footage. Projects in which 50% or less of the building's leasable square footage is occupied by an owner or tenant should pursue LEED for Core & Shell certification.

When to Use LEED for Schools

LEED for Schools addresses design and construction activities for both new school buildings and major renovations of existing school buildings.

- The activities must be performed in the construction or major renovation of an academic building on K–12 school grounds.

- Other projects on a school campus may qualify under two or more LEED rating system project scopes; for example, nonacademic buildings on a school campus, such as administrative offices, maintenance facilities, or dormitories, are eligible for either LEED for New Construction and Major Renovation or LEED for Schools.

- Projects involving postsecondary academic buildings or prekindergarten buildings may choose to use either LEED for New Construction and Major Renovation or LEED for Schools.

When to Use LEED for Healthcare

The LEED for Healthcare Green Building Rating System was developed to meet the unique needs of the healthcare market, including the following:

- In-patient care facilities, licensed out-patient care facilities, and licensed long-term care facilities, as well as medical offices, assisted-living facilities, and medical education and research centers.

- LEED for Healthcare addresses issues such as increased sensitivity to chemicals and pollutants, traveling distances from parking facilities, and access to natural spaces.

When to Use LEED for Core & Shell

The LEED for Core & Shell Rating System is a market-specific application developed to serve the speculative development market, in which project teams do not control all scopes of a whole building's design and construction.

- Buildings such as commercial office buildings, medical office buildings, retail centers, warehouses, and lab facilities can qualify.

- Depending on how the project is structured, this scope can vary significantly from project to project.

- The rating system addresses a variety of project types and a broad project range.

- The rating system can be used for projects in which the developer controls the design and construction of the entire core and shell base building (such as the mechanical, electrical, plumbing, and fire protection systems), but has no control over the design and construction of the tenant fit-out.

- The owner must occupy 50% or less of the building's leasable square footage. Projects in which more than 50% of the building's tenant space is occupied by an owner should pursue LEED for New Construction and Major Renovation certification.

- Because there are many unknowns in a core and shell project type (i.e., who the tenant will be, what the occupancy rate of the future tenant will be, etc.), the LEED Core & Shell Rating System provides guidance & procedures to follow, as well as default figures that must be used, these include:

  - Default Occupancy Counts;

  - Core & Shell Energy Modeling Guidelines;

  - Core & Shell Project Scope;

  - Tenant Lease or Sales Agreement; and

  - LEED for Core & Shell Precertification Guidance.

When to Use LEED for Commercial Interiors

LEED for Commercial Interiors addresses the specifics of tenant spaces primarily in office, retail, and institutional buildings.

- Tenants who lease their space or do not occupy the entire building are eligible.

- The system is designed to work hand in hand with the LEED for Core & Shell certification system.

- Tenant spaces that are not within a LEED for Core & Shell building are also eligible for certification under LEED for Commercial Interiors.

When to Use LEED for Retail

LEED for Retail: New Construction 2009 and LEED for Retail: Commercial Interiors 2009 recognize the unique nature of the retail environment and address the different types of spaces retailers need for their distinctive product lines.

- LEED for Retail: New Construction allows for the whole-building certification of freestanding retail buildings.

- LEED for Retail: Commercial Interiors allows tenants to certify their build-out, regardless of their control over the building envelope.

- Existing freestanding retailers can green their real estate portfolio through LEED for Existing Buildings: Operations & Maintenance.

When to Use LEED for Existing Buildings: Operations & Maintenance

LEED for Existing Buildings: Operations & Maintenance was designed to certify the sustainability of the ongoing operations of existing commercial and institutional buildings.

- All commercial and institutional buildings, including offices, retail and service establishments, libraries, schools, museums, churches, and hotels, as well as residential buildings of four or more habitable stories are eligible.

- The rating system encourages owners and operators of existing buildings to implement sustainable practices and reduce the environmental impacts of their buildings over their functional life cycles.

- The rating system addresses exterior building site maintenance programs, water and energy use, environmentally preferred products and practices for cleaning and alterations, sustainable purchasing policies, waste-stream management, and ongoing indoor environmental quality.

- The rating system is targeted to single buildings, whether owner-occupied, multi-tenanted, or multiple-building campus projects. If there are multiple buildings on the same campus, each must certify individually. Also, It is a whole-building rating system; individual tenant spaces are ineligible.

When to Use LEED for Homes

Any project that participates in LEED for Homes must be defined as a "dwelling unit" by all applicable codes. This requirement includes, but is not limited to, the following conditions:

- A dwelling unit must include "permanent provisions for living, sleeping, eating, cooking, and sanitation." Every participating home must have a cooking area and a bathroom.

- The rating system focuses specifically on single-family and small multi-family homes.

- Single-family homes and multi-family buildings up to three stories can be certified under LEED for Homes. If there are more than three stories, projects may use LEED NC, or the LEED for Homes Mid-Rise Pilot.

When to Use LEED for Neighborhood Development

LEED for Neighborhood Development has been designed to be the first national standard for neighborhood design and mixed-use communities. Focus areas include the following:

- The principles of smart growth, new urbanism best practices, and green building.

- The land-use planning of an entire neighborhood, including buildings, infrastructure, street design, and open space.

- Current participants are in areas with significant previous development and existing infrastructure, as well as areas with more compact development forms where alternative transportation modes, such as transit or walking, are viable.

- Eligible projects successfully protect and enhance the overall health, natural environment, and quality of life of our communities.

- A large variety of project sizes is eligible. However, the project must include a residential component (new or existing).

LEED Online

LEED Online is the required resource that project teams with registered projects use to manage the LEED certification process. With LEED Online, you can:

- Manage project details;
- Complete documentation requirements for LEED credits and prerequisites;
- Upload supporting files;
- Submit applications for review;
- Receive reviewer feedback; and
- Ultimately earn LEED certification.

The Online system provides status updates and time lines for your LEED project team. It is a user-friendly system that walks you through all the steps in the review and certification process. Within LEED Online, you'll find the required submittal templates and other LEED management resources.

Project Registration and Certification Process

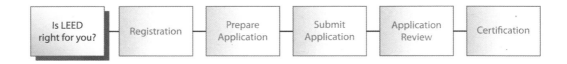

Registration

Registering for LEED certification entails two steps:

1. Fill out an online registration form at the LEED Online website, https://www.gbci. org/DisplayPage.aspx?CMSPageID=174. You will be required to enter basic project details, which can be edited at a later time if they change.

2. You will be required to pay a flat registration fee up front at the time of registration. This fee is discounted for USGBC members.

Certification

Obtaining LEED certification entails the following steps:

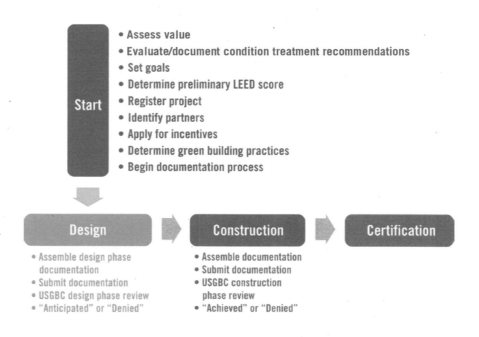

Certification Fee

The LEED certification fee is based on the rating system that the project is certifying under and the size of the project. The fee is paid when the project team submits documentation for review via LEED Online. Once a project team submits design and construction phase documentation to LEED Online, the following, third-party certification process takes place:

1. **Preliminary Review:** Complete your LEED Online documentation and submit the credit/prerequisite for review. Whether or not the credit/prerequisite is a design or construction phase credit, it will undergo a preliminary review first. If the credit/prerequisite is awarded at this stage, you are done! If it isn't awarded and there are questions or missing documents, the project team will receive comments and clarification requests back after the preliminary review.

2. **Final Review:** This is when the project team will have the chance to directly address the comments from the preliminary review. You can submit new documents and provide a response in order to demonstrate that your team has achieved the prerequisite or credit.

3. **Appeal Review:** If you haven't been awarded a credit after the preliminary and final reviews, you have another chance to achieve it and submit new documentation in the appeal review. There is an additional fee associated with the appeal review.

Exceptions

LEED for Homes and LEED for Neighborhood Development have different certification processes from the other rating systems. LEED for Homes has five steps and different project team players. LEED for Neighborhood Development has three, broad-reaching steps that depend on the actual neighborhood development processes, which can take years or decades to complete:

1. The review occurs prior to completion of the permitting process.

2. The certification applies to an approved development plan.

3. The review encompasses the completed neighborhood development.

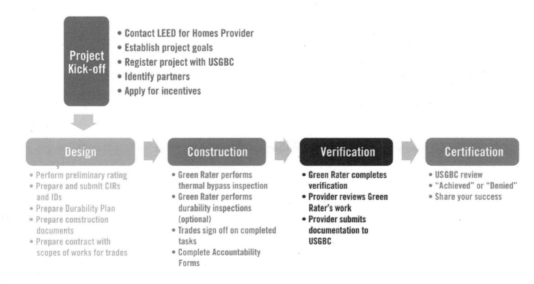

Credit Interpretation Requests

A Credit Interpretation Request, or CIR, is a formal question asked of GBCI from the project team. GBCI responds with a Credit Interpretation Ruling. Teams submit a CIR when they require clarification on their approach to achieving a specific prerequisite or credit. There is a fee for submitting a CIR; details for submitting CIRs are available online at www.gbci.org.

Documentation Guidelines

When you submit a project for LEED certification, you are required to submit some general documents. These are universal requirements meant to provide an explanation of your project scope and highlights:

- Project narrative: The team should submit a short narrative (one to three pages) that describes the background/history of the project, the use(s) of the building in some detail, the location and surrounding area of the building, and any other project attributes the team would like to highlight.

- Project photographs or renderings.

- Elevations.

- Typical floor plans.

- Project details also must be provided and applied consistently across all LEED credits. These include the gross square footage of the building and the number of occupants.

- It is important to distinguish between the project boundary, the LEED project boundary, and the property boundary:
  - **The project boundary** is the platted property line of the project defining land and water within it. Projects located on publicly owned campuses that do not have internal property lines shall delineate a sphere-of-influence line to be used in place of a property line. The phrase "project site" is equivalent to the land and water inside the project boundary. The project may contain noncontiguous parcels, as long as their perimeters are no farther apart than a quarter-mile walk, and each separate parcel must independently comply with rating system requirements. Projects may also have enclaves of nonproject properties that are not subject to the rating system, but such enclaves cannot exceed 2% of the total project area and cannot be described as certified.
  - **The LEED boundary** is the portion of the project site submitted for LEED certification. For single-building developments, this is the entire project scope and is generally limited to the site boundary. For multiple-building developments, the LEED project boundary may be a portion of the development as determined by the project team.
  - **The property boundary and/or property area** is the total area within the legal property boundaries of a site; it encompasses all areas of the site, including constructed and nonconstructed areas.

LEED Professionals

There are three tiers of LEED Professional Credentials which distinguish professionals with basic, advanced, and extraordinary levels of knowledge:

- ● LEED Green Associate;
- ● LEED Accredited Professional with specialty; and
- ● LEED Fellow.

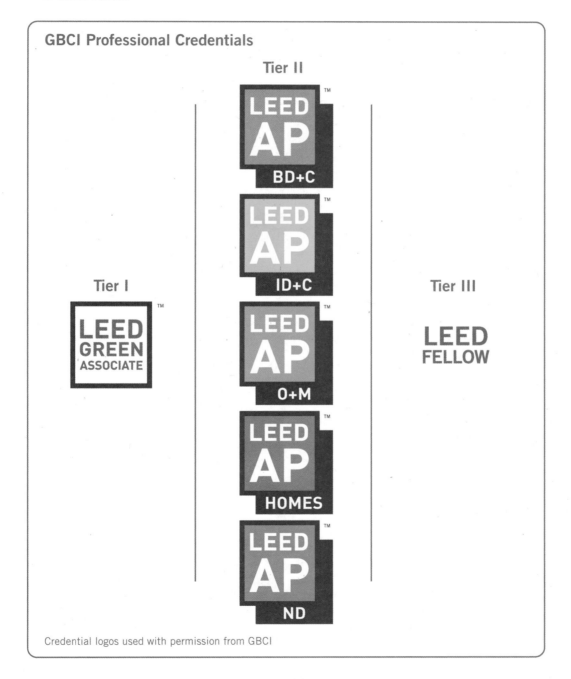

GBCI Professional Credentials

Tier I — LEED GREEN ASSOCIATE™

Tier II — LEED AP™ BD+C, LEED AP™ ID+C, LEED AP™ O+M, LEED AP™ HOMES, LEED AP™ ND

Tier III — LEED FELLOW

Credential logos used with permission from GBCI

LEED Green Associate

What:
A Green Associate credential attests to the candidate's knowledge of good environmental practice and skill, and reflects an understanding and support of green design, construction, and operations. There are no specialties within the LEED Green Associate tier. The Green Associate credential also serves as the first step for professionals pursuing a LEED Accredited Professional specialization.

How:
LEED Professional Credentialing, Tier I:

- You must agree to the Disciplinary Policy and Credential Maintenance Requirements as outlined at www.gbci.org.

- You must document your involvement in support of LEED OR be employed in a sustainable field of work OR be engaged in an education program in green building principles and LEED.

- You should submit to an application audit; 5% to 7% of all applications will be audited. You will be notified immediately if you are chosen for an audit and will be notified of your eligibility within 14 days.

- Candidates who successfully pass the LEED Green Associate Exam must complete 15 continuing education (CE) hours biennially.

LEED Accredited Professional With Specialty

What:
A LEED Accredited Professional (AP) with specialty credential signifies an advanced depth of knowledge of green building practices. The Green Associate Exam is Part 1 of the LEED AP with specialty exams; if you passed and maintained your Green Associate credential, you will not have to take that exam again. LEED APs will earn their credentials in one (or more) of the following five specialties:

- Building Design + Construction (BD+C);
- Operations + Maintenance (O+M):
- Interior Design + Construction (ID+C);
- Homes; and/or
- Neighborhood Development (ND).

How:
LEED Professional Credentialing, Tier II:

- You must agree to the Disciplinary Policy and Credential Maintenance Requirements as outlined at www.gbci.org.

- You must document your involvement in support of LEED OR be employed in a sustainable field of work OR be engaged in an education program in green building principles and LEED.

- You must have previous experience with a LEED Registered Project within three years of your application submittal date.

- Candidates who successfully pass the LEED AP specialty exam must complete 30 CE hours biennially.

LEED Fellow

What:

This level is currently under development.

How:

LEED Professional Credentialing, Tier III (currently under development):

- This tier will distinguish an elite class of leading professionals.

- Fellows will contribute to the standards of practice and body of knowledge for achieving continuous improvement in the green building field.

Why Seek LEED Credentials?

Individual benefits:

- Provides a marketable credential to an employer, prospective employer, or client;

- Provides a listing on the GBCI website directory of LEED Professionals;

- Awards a LEED Professional certificate; and

- Recognizes the individual for involvement in the LEED certification process.

Employer benefits:

- Establishes eligibility for projects on which owners are mandating the participation of a LEED AP;

- Strengthens qualifications when responding to requests for proposals (RFPs) requiring a LEED AP; and

- Encourages the growth of knowledge and understanding of the LEED certification process.

Industry benefits:

- Encourages and promotes a higher understanding of LEED; and

- Supports and facilitates transformation of the built environment.

LEED Professional Ethics

Once you have your LEED credential, you also have a responsibility! GBCI created a Disciplinary Policy that serves as a code of conduct for individuals seeking certification. The Disciplinary Policy also establishes a fair process for addressing noncompliance. Matters are investigated by a Disciplinary Review Committee and presented for judgment before a Disciplinary Hearing Committee. These committees operate independently of each other. The GBCI Credential Steering Committee is available to hear appeals of Disciplinary Hearing Committee decisions and is the final decision maker on behalf of GBCI.

The general principles of the Disciplinary Policy require that LEED-certified individuals must:

- Be truthful, forthcoming, and cooperative in their dealings with GBCI;
- Be in continuous compliance with GBCI rules (as amended from time to time by GBCI);
- Respect GBCI intellectual property rights;
- Abide by laws related to the profession and to general public health and safety; and
- Carry out their professional work in a competent and objective manner.

USGBC CATEGORY REVIEW

Explain what USGBC and GBCI do. How do they interact?

What are the differences between a prerequisite and a credit?

What categories of concern do the LEED rating systems address?

LEARNING ACTIVITIES

INVESTIGATE

Look around your community. Identify as many examples as you can of projects that meet the criteria to be rated under each of the LEED rating systems. (They don't have to be projects green enough to be certified, just that are of the right type for the rating system.)

LEED for New Construction and Major Renovation
LEED for Existing Buildings: Operations & Maintenance
LEED for Commercial Interiors
LEED for Core & Shell
LEED for Homes
LEED for Schools
LEED for Retail
LEED for Healthcare
LEED for Neighborhood Development

TRY IT OUT

Review three or more credits in a LEED reference guide (or from the appendix in this study guide). Identify each credit component.

DISCUSS IN A GROUP

Divide into three groups: one each for the economy, the environment, and social issues. Consider the following events and evaluate within your small groups how your topic area is affected:

- A new power plant opens in your town.
- Older housing is demolished to make way for a new condo tower.
- A developer purchases 120 acres of open space for a new retail center.
- A new bus route is added to serve your neighborhood.

WATCH IT

Watch the LEED Version 3 Online demo (a five-minute video) on your computer at www.gbci.org. You may also view the video on a large screen at www.vimeo.com (search for "LEED v3 Online Demo").

TRY IT OUT

Provide an aerial photo of a site. Draw a LEED boundary and provide an explanation for the area selected.

PRACTICE QUESTIONS

1. How many levels of LEED Professional Accreditation are available?

a.) 6
b.) 4
c.) 1
d.) 3

2. Which of the following is a primary responsibility of the U.S. Green Building Council?

a.) Developing the LEED Professional Accreditation exams
b.) Administering LEED project reviews and certification processes
c.) Establishing continuing education requirements for LEED Accredited Professionals
d.) Providing and developing LEED-based education and research programs

3. A development company is designing a seven-story, 100,000-square-foot condominium building. The developers will be responsible for completing the interior finishes, but will not be supplying furniture or appliances. What LEED rating system would be most relevant for this project type?

a.) LEED for New Construction and Major Renovation
b.) LEED for Homes
c.) LEED for Commercial Interiors
d.) LEED for Core & Shell

4. The carbon overlay in LEED is used for what purpose?

a.) To prioritize the relative impact of credits on greenhouse gas emissions
b.) To identify the major contributors to environmental degradation
c.) To quantify the relative impact of different energy efficiency measures
d.) To rank the feasibility of various green building strategies

5. What is the procedure required to achieve LEED certification?

a.) Register a project with GBCI, pay applicable review fees, and submit documentation.
b.) Retain a LEED professional, register the project, and pay applicable fees.
c.) Submit documentation, obtain a preliminary rating, and pay applicable certification fees.
d.) Register the project with GBCI, demonstrate environmental innovation, and pay applicable fees.

6. What is the earliest point at which a LEED for Schools project can be certified?

a.) After one year of occupancy and after all commissioning activities are complete.
b.) After the project team has registered and submitted all project specifications.
c.) After final punch-list items are complete and all review fees are paid.
d.) After building completion and once all submittals and clarifications are reviewed.

Answer Key on Page 133

PRACTICE QUESTIONS

7. The licensed-professional exemption is used by a project team to do what?

a.) Achieve continuing education credit for primary team members.

b.) Capture federally available tax credits for the project.

c.) Bypass otherwise required submittals.

d.) Streamline the permitting process in many jurisdictions.

8. A team is unclear whether a proposed project strategy will achieve a specific LEED credit the team is pursuing. The team decides it should submit a Credit Interpretation Request (CIR). Prior to submitting the CIR for review, which strategies should the project team consider (select three)?

a.) Review the credit intent and self-evaluate whether the strategy meets that intent.

b.) Contact LEED customer service to determine whether the CIR is likely to be successful.

c.) Review past CIRs to see whether this issue has been addressed in the past.

d.) Consult the appropriate LEED reference guide for a more detailed explanation.

e.) Contact its local chapter to receive a preliminary ruling.

f.) Identify other issues to address within the CIR.

Answer Key on Page 134

KEY TERMS TO KNOW:

Cover up the left side of the page and test yourself to see if you can summarize the definitions of the following key terms. Or better yet, make flash cards.

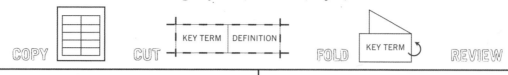

COPY CUT | KEY TERM | DEFINITION | FOLD | KEY TERM | REVIEW

LEED Credit	An optional LEED Green Building Rating System™ component whose achievement results in the earning of points toward certification.
LEED Credit Interpretation Request	A formal USGBC process in which a project team experiencing difficulties in the application of a LEED prerequisite or credit can seek and receive clarification, issued as a **Credit Interpretation Ruling**. Typically, difficulties arise when specific issues are not directly addressed by LEED reference guides or a conflict between credit requirements arises.
LEED Intent	The primary goal of each prerequisite or credit.
LEED® Green Building Rating System™	A voluntary, consensus-based, market-driven building rating system based on existing, proven technology. The LEED Green Building Rating System™ represents USGBC's effort to provide a national benchmark for green buildings. Through its use as a design guideline and third-party certification tool, the LEED Green Building Rating System aims to improve occupant well-being, environmental performance, and economic returns using established and innovative practices, standards, and technologies.
LEED Prerequisite	A required LEED Green Building Rating System™ component whose achievement is mandatory and does not earn any points.

LEED Technical Advisory Group (TAG)	A committee consisting of industry experts who assist in interpreting credits and developing technical improvements to the LEED Green Building Rating System™.

WHAT ABOUT SUSTAINABLE SITES?

- Can site design save energy?
- Can site design reduce surface-water pollution?
- Can site design create a safer, more comfortable environment?
- How does site design contribute to personal mobility?
- How does site design contribute to the protection of habitat or agricultural resources?
- Can we create a building that restores natural runoff patterns?
- Can we create a building that helps restore an urban fabric?

Read pages 50-59 of the Green Building and LEED Core Concepts Guide, Second Edition, or pages 25-36 in the First Edition.

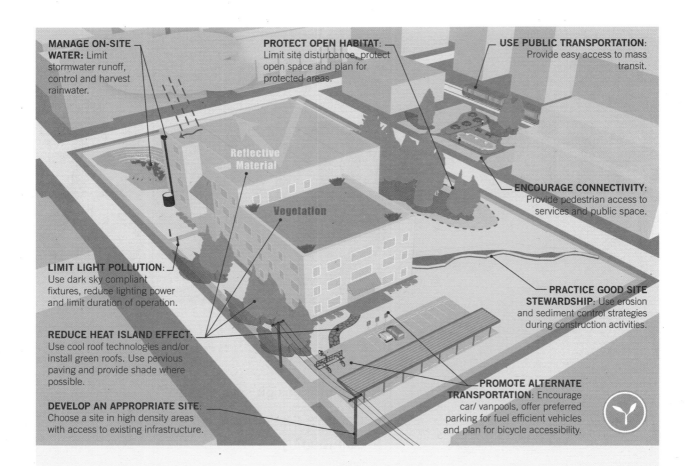

MANAGE ON-SITE WATER: Limit stormwater runoff, control and harvest rainwater.

PROTECT OPEN HABITAT: Limit site disturbance, protect open space and plan for protected areas.

USE PUBLIC TRANSPORTATION: Provide easy access to mass transit.

ENCOURAGE CONNECTIVITY: Provide pedestrian access to services and public space.

LIMIT LIGHT POLLUTION: Use dark sky compliant fixtures, reduce lighting power and limit duration of operation.

PRACTICE GOOD SITE STEWARDSHIP: Use erosion and sediment control strategies during construction activities.

REDUCE HEAT ISLAND EFFECT: Use cool roof technologies and/or install green roofs. Use pervious paving and provide shade where possible.

DEVELOP AN APPROPRIATE SITE: Choose a site in high density areas with access to existing infrastructure.

PROMOTE ALTERNATE TRANSPORTATION: Encourage car/ vanpools, offer preferred parking for fuel efficient vehicles and plan for bicycle accessibility.

Reflective Material

Vegetation

SUSTAINABLE SITES OVERVIEW

In this chapter, you will learn that a project's location is the foundation for sustainability. The selection and development of a site has a dramatic impact on the performance of a building over the course of its life, from the way people travel to the building to how the project coexists with the local ecosystem. Four key issues help define how a project's location affects the sustainability of the project over its lifetime:

- **Transportation;**
- **Site selection;**
- **Site design and management; and**
- **Stormwater management.**

Transportation

Buildings and the way we develop land generate demand for transportation. Transportation has profound impacts on the environment, the economy, and the community.

Environment: Vehicle emissions are one of the leading contributors to climate change, smog, acid rain, and other air quality problems. Moreover, miles of roads and parking lots not only affect water quality but also require massive amounts of energy to construct.

Economy: Beyond the environmental cost of vehicle transportation lies a huge economic cost to build and maintain roads. The economic cost of purchasing gas and oil to fuel these vehicles represents a significant portion of a household's expenditures, ranking with electricity and clothing.

Community: Human health is affected by transportation in numerous ways, including climate change and air quality issues. But perhaps most troublesome is the reduction in physical activity that has come with the increased use of vehicles. With the incidence of obesity and overweight individuals rising at alarming rates in this country, there are plenty of data to support the value of getting out of the car and walking or biking instead.

Strategies

Now that we know the impacts of transportation, the question is, how can the selection of a site change these trends? Making it easy for people to get places, either without driving at all or without driving alone, thereby reducing the number of vehicles on the road, is a key priority for green projects. There are many ways developers and design teams can encourage this behavior, through simple and intentional steps.

Photo by Kalpana Kuttaiah

Strategies

Locate the project near mass transit.	Make it easy on employees to take the bus or train to work by picking a site that is within easy walking distance of bus or train stops.
Limit parking.	You can't drive there if there's nowhere to park, and that's the point! Challenge the perception of how much parking is really needed to support the project, and provide the least amount possible.

Strategies

Encourage carpooling.

Give the best parking spots to those who carpool, as a token of your appreciation.

Promote alternative-fuel vehicles.

Provide the best parking spots for alternative-fuel vehicles. Take the next step by providing electric-car charging stations. Go even further by actually providing an alternative-fuel vehicle for building occupants' usage.

Photo by Shawn Hamlin

Offer incentives.

Nothing speaks louder than cash. Reward employees or shoppers who arrive by bike.

Support alternative transportation.

Help organize building occupants so they know their options for commuting. Provide a bulletin board with carpool groups and bus schedules.

Site Selection

Buildings and the way we develop land affect ecosystems in a number of ways and can seriously harm existing wildlife habitats.

Environment: Destruction of wildlife habitats for new development displaces native animals, threatening their ability to survive. The loss of land over time to development lessens the biodiversity of both plant and animal species.

Economy: Appropriate development of sites away from critical natural habitats can also help protect buildings from natural disasters such as landslides and floods, which are extremely costly to project owners.

Community: Directing development through intentional planning preserves natural areas for enjoyment and recreation. Protecting areas with diverse plant and animal species not only benefits wildlife, but also provides humans the needed opportunity to connect with their natural surroundings.

Strategies

A sustainable site is one that is located within or near existing development to leverage existing infrastructure and provide basic services to building occupants. Additionally, it prioritizes potential sites that have been previously developed, minimizing the degradation of greenfield sites.

Selecting sites based on their environmental attributes is a way to significantly minimize a project's impact on the environment. Focus on elements such as those listed in the chart below.

Before Site Redvelopment.

After Site Redevelopment.
Photos courtesy of Urban-Advantage.com

Strategies

Increase density.	Instead of a sprawling one-story development, consider building up instead of out. This will maximize the square footage inside your building while minimizing the amount of land that is being used.
Choose redevelopment.	Old buildings offer more than history. They are often well located, with access to existing services and public transportation. Reuse is one of the single most important concepts of sustainability.
Protect habitat.	Leave undeveloped areas with critical habitats undamaged and undeveloped. Even within a site, consider which portions of it might be best left as open space and focus density in the areas where it makes sense.

Site Design and Management

Location isn't everything; even if you've picked the perfect site, there are still a lot of issues to consider. Projects that are plunked down on a site without any regard for the local environment can tax local resources and wreak havoc on their surroundings.

Environment: The introduction of nonnative plant species can require irrigation and chemicals, both of which threaten the quantity and quality of available water.

Economy: Maintaining plant species that are not well adapted to an area is costly in terms of watering and the use of fertilizers and pesticides.

Community: Place matters, and people seek connections with the natural environment. When projects ignore the context around them, they slowly degrade what makes a place special.

Strategies

Sustainable site design and management involves the selection of native and/or adaptive plantings, water-efficient irrigation systems, minimal hardscape, and optimal exterior lighting design. A sustainable landscape entails not only designing and creating, but also maintaining the landscape in a sustainable manner.

Sustainable landscapes are those that reduce environmental impacts, minimize maintenance costs, and contribute to the restoration and regeneration of an area.

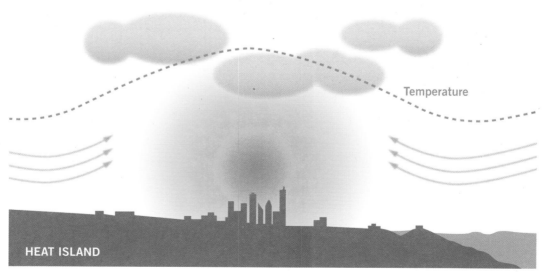

Heat Island Effect is caused by the absorption of heat by hardscapes, such as dark, nonreflective pavement and buildings, and its radiation to surrounding areas.

Strategies

Build small.

Think small; small buildings require less resources than large buildings. Efficient spaces are smart and meet people's needs without being excessive.

Minimize hardscape.

Water flows right over hard surfaces like paved parking lots, carrying away a precious resource that is now filled with pollutants. Consider using materials that allow water to flow through them, or use the least amount of hard space you can.

Minimize water usage.

Not every building needs a green lawn in front of it. Landscape designs that use plants that grow naturally in an area add to people's sense of place and cut down on the amount of water needed to keep the landscape looking good.

Photo by Adrain Velicescu

Use reflective materials.

Ever run barefoot across black pavement to get to your car in the summer? Black absorbs heat, but light colors reflect heat back into the atmosphere. Instead of using a black roof or asphalt, use light-colored materials so that the overall temperature of an area doesn't rise.

Develop a sustainable management plan.

There are lots of people involved in taking care of a building. Don't leave it up to chance that everyone will understand the importance of green. Write a plan for taking care of the building, and share it with those involved in the maintenance of the facility.

Reduce light pollution.

Gazing up at the night sky in a major metropolitan area is nothing like pulling over on a country road. Lighting from buildings pollutes the night sky, inhibiting not only star gazers but also nocturnal wildlife. Make sure that buildings are lit minimally at night and that this lighting is directed downward rather than up into the sky.

Photo by ©Christian Richters

Stormwater Management

The fastest-growing source of surface-water degradation is the expansion of impervious surfaces. This is due to both a decrease in filtration and the buildup on hardscape areas of contaminants that are concentrated in surface runoff during heavy rainfall.

Increased urbanization and development lead to more hardscapes and impervious surfaces. These, in turn, increase stormwater runoff which can accelerate erosion, thereby carrying particulates and chemicals into nearby water bodies, such as rivers and streams. Ultimately, this can jeopardize water quality, aquatic life, and recreation opportunities in nearby rivers and streams (which often provide local water supplies).

Environment: Development frequently disturbs a site's natural ecological system. The loss of existing plants and animals can be devastating, because they provide important ecological services, including effective management of stormwater.

Economy: The cost of providing man-made infrastructure to handle stormwater is significant. Such systems carry huge up-front installation costs as well as ongoing maintenance costs.

Community: Surface water provides recreational opportunities in the form of swimming, fishing, and other activities, but only if it's clean. Surface-water quality can be improved by properly managing stormwater runoff.

Strategies

Outdoor Filtration System, Photo by Hillary Platt

Minimize impervious areas.

Think of ways to get stormwater to stay on your site rather than directing it away. Green roofs and pervious pavers are good ways to get water to infiltrate a site.

Photo courtesy of Green Building Services

Control stormwater.

Another way to improve the quality of water leaving your site is to slow it down. Rain gardens and bioswales are simply technical terms for landscape features that are designed to capture and slow down water leaving the site after a big storm. Nature knows best, and plants essentially clean the water of pollutants.

Photo by Kalpana Kuttaiah

Harvest rainwater.

What better way to respect the value of rain than by saving it? Depending on the size of your project, harvesting rainwater can be as simple as connecting a barrel to a downspout or as advanced as tying the collected rainwater into the plumbing system to provide water to flush toilets.

SS CATEGORY REVIEW

Why is it beneficial to develop in high-density areas and use existing infrastructure for a project site?

When considering community connectivity, we refer to basic services. What are these?

What should a sustainable management plan address for building landscapes and hardscapes?

Why do we care about light pollution?

How can you minimize impervious areas on a project site?

LEARNING ACTIVITIES

PLAY A GAME

Go around the room and play the alphabet game. Starting with the first person, have him or her name a sustainable sites strategy, issue, or idea that starts with the letter "A" and offer a brief explanation. For example, the first person might say, "Automobiles—single drivers in automobiles is one of the leading contributors to climate change." The next person takes "B," and so on.

INVESTIGATE

Consider two sites in your community (these could be your home and place of work). What are their pluses and minuses as sustainable sites?

STRATEGY	ADVANTAGES	CHALLENGES
Access to mass transit		
Support for alternative transportation		
Minimized hardscape		
Minimized water usage		
Reflective materials		
Stormwater control		
Rainwater harvesting		
Site lighting		
Heat island effect		

GROUP ACTIVITY

Find three sample site photos or images. Split into three small groups, with each group taking one of the sites. Within the groups, list the strategies being used at the site in the areas of transportation, site selection, site design and management, and stormwater management. Where are the opportunities? Come back together as one large group and share your findings.

PRACTICE QUESTIONS

1. **What metric is the best indicator of transportation impacts associated with a building project?**

   a.) Street grid density
   b.) Availability of public transportation
   c.) Vehicle miles traveled
   d.) Parking capacity

2. **Decreasing impervious surfaces on a project site will ___?**

   a.) Decrease percolation rates
   b.) Reduce potable water usage
   c.) Reduce stormwater runoff
   d.) Eliminate sewage piping

3. **What is acknowledged as one of the greatest threats to surface-water quality?**

   a.) Draining of local aquifers
   b.) Recreational boating
   c.) Rainwater harvesting
   d.) Nonpoint-source pollution

4. **A project that specifies exterior surfaces with high solar reflectance index (SRI) values is contributing to which environmental benefit?**

   a.) Reduced heat island effect
   b.) Support for renewable energy
   c.) Protection of the dark-sky initiative
   d.) Improved stormwater quality

5. **A project is in the pre-design phase and the site has already been selected. The team wants to increase the open space on the project site. Which strategies should it consider (select two)?**

   a.) Increase the floor-to-area ratio of the building.
   b.) Implement a construction activity pollution plan.
   c.) Use pervious paving materials for the parking area.
   d.) Locate parking underground.
   e.) Select drought-tolerant plantings.

6. **An industrial facility is located in an area with no public transportation. In which ways can the project team reduce the project's transportation impact (select two)?**

   a.) Provide a carpooling incentive to building occupants.
   b.) Upgrade the company cars to hybrids.
   c.) Install solar panels to power parking lot lighting.
   d.) Provide occupants a fuel subsidy.

Answer Key on Page 134

PRACTICE QUESTIONS

7. A project adjacent to protected forestland that is home to a variety of plant and animal life wants to reduce the impact of its site lighting. To achieve this, the project team installs exterior lighting that ___?

a.) Provides for tasteful, decorative appearance.

b.) Reduces the need for night-time security.

c.) Adequately illuminates the night sky.

d.) Does not trespass onto adjacent properties.

8. Prior to final selection of the project site, the owner and design team should confirm that the site is ___?

a.) Compliant with the green design criteria

b.) Compliant with the sustainable building codes

c.) Previously undeveloped

d.) Removed from other development

Answer Key on Page 134

KEY TERMS TO KNOW:

Cover up the left side of the page and test yourself to see if you can summarize the definitions of the following key terms. Or better yet, make flash cards.

COPY   CUT | KEY TERM | DEFINITION | FOLD   KEY TERM   REVIEW

Key Term	Definition
Acid Rain	The precipitation of dilute solutions of strong mineral acids, formed by the mixing in the atmosphere of various industrial pollutants (primarily sulfur dioxide and nitrogen oxides) with naturally occurring oxygen and water vapor.
Alternative Fuel Vehicles	Vehicles that use low-polluting, nongasoline fuels, such as electricity, hydrogen, propane or compressed natural gas, liquid natural gas, methanol, and ethanol. In LEED, efficient gas-electric hybrid vehicles are included in this group.
Biodiversity	The variety of life in all forms, levels, and combinations, including ecosystem diversity, species diversity, and genetic diversity.
Biomass	Plant material from trees, grasses, or crops that can be converted to heat energy to produce electricity.
Bioswale	A stormwater control feature that uses a combination of an engineered basin, soils, and vegetation to slow and detain stormwater, increase groundwater recharge, and reduce peak stormwater runoff.

Brownfield	Previously used or developed land that may be contaminated with hazardous waste or pollution. Once any environmental damage has been remediated, the land can be reused. Redevelopment on brownfields provides an important opportunity to restore degraded urban land while promoting infill and reducing sprawl.
Building Density	The floor area of the building divided by the total area of the site (square feet per acre).
Building Footprint	The area on a project site that is used by the building structure, defined by the perimeter of the building plan. Parking lots, landscapes, and other nonbuilding facilities are not included in the building footprint.
Community Connectivity	The amount of connection between a site and the surrounding community, measured by proximity of the site to homes, schools, parks, stores, restaurants, medical facilities, and other services and amenities.
Development Density	The total square footage of all buildings within a particular area, measured in square feet per acre or units per acre.
Diversity of Uses or Housing Types	The number of types of spaces or housing types per acre. A neighborhood that includes a diversity of uses—offices, homes, schools, parks, stores—encourages walking, and its residents and visitors are less dependent on personal vehicles. A diversity of housing types allows households of different types, sizes, ages, and incomes to live in the same neighborhood.

Dry Ponds	Excavated areas that detain stormwater and slow runoff but are dry between rain events. Wet ponds serve a similar function but are designed to hold water all the time.
Floodplain	Land that is likely to be flooded by a storm of a given size (e.g., A 100-year storm).
Floor-To-Area Ratio	The relationship between the total building floor area and the allowable land area the building can cover. In green building, the objective is to build up rather than out because a smaller footprint means less disruption of the existing or created landscape.
Foot Candle	A measure of the amount of illumination falling on a surface. A footcandle is equal to one lumen per square foot. Minimizing the number of footcandles of site lighting helps reduce light pollution and protect dark skies and nocturnal animals.
Heat Island Effect	The absorption of heat by hardscapes, such as dark, nonreflective pavement and buildings, and its radiation to surrounding areas. Particularly in urban areas, other sources may include vehicle exhaust, air-conditioners, and street equipment; reduced airflow from tall buildings and narrow streets exacerbates the effect.
Imperviousness	The resistance of a material to penetration by a liquid. The total imperviousness of a surface, such as paving, is expressed as a percentage of total land area that does not allow moisture penetration. Impervious surfaces prevent rainwater from infiltrating into the ground, thereby increasing runoff, reducing groundwater recharge, and degrading surface water quality.

Native and Adapted Plants	Native plants occur naturally in a given location and ecosystem. Adapted plants are not native to a location but grow reliably with minimal attention from humans. Using native and adapted plants can reduce the amount of water required for irrigation, as well as the need for pesticides or fertilizers, and may provide benefits for local wildlife. Native plants are considered low maintenance and not invasive.
Perviousness	The percentage of the surface area of a paving material that is open and allows moisture to pass through the material and soak into the ground below.
Prime Farmland	Previously undeveloped land with soil suitable for cultivation. Avoiding development on prime farmland helps protect agricultural lands, which are needed for food production.
Rain Garden	A stormwater management feature consisting of an excavated depression and vegetation that collects and filters runoff and reduce peak discharge rates.
Site Disturbance	The amount of a site that is disturbed by construction activity. On undeveloped sites, limiting the amount and boundary of site disturbance can protect surrounding habitat.
Solar Reflectance Index (SRI)	A measure of how well a material rejects solar heat; the index ranges from 0 (least reflective) to 100 (most reflective). Using "cooler" materials helps prevent the urban heat island effect (the absorption of heat by roofs and pavement and its radiation to the ambient air) and minimizes demand for cooling of nearby buildings.

Stormwater Runoff	Water from precipitation that flows over surfaces into sewer systems or receiving water bodies. All precipitation that leaves project site boundaries on the surface is considered stormwater runoff.
Street Grid Density	An indicator of neighborhood density, calculated as the number of centerline miles per square mile. Centerline miles are the length of a road down its center. A community with high street grid density and narrow, interconnected streets is more likely to be pedestrian friendly than one with a low street grid density and wide streets.
Transportation Demand Management	The process of reducing peak-period vehicle trips.
Vehicle Miles Traveled (VMT)	A measure of transportation demand that estimates the travel miles associated with a project, most often for single-passenger cars. LEED sometimes uses a complementary metric for alternative-mode miles (e.g., In high-occupancy autos).
Wetland Vegetation	Plants that require saturated soils to survive or can tolerate prolonged wet soil conditions.
Xeriscaping	A landscaping method that makes routine irrigation unnecessary by using drought-adaptable and low-water plants, as well as soil amendments such as compost and mulches to reduce evaporation.

WHAT ABOUT WATER EFFICIENCY?

- Where does drinking water come from?
- What fixtures or appliances use the most water in the building?
- Can we have a "zero net water" building? If not, how close can we come?
- Can you trace where water goes after it lands on your roof?
- What's the difference between a low-performing and a high-performing building when it comes to water?
- What is the difference between a storm drain and a sewer?

Read pages 60-63 of the Green Building and LEED Core Concepts Guide, Second Edition, or pages 37-42 in the First Edition.

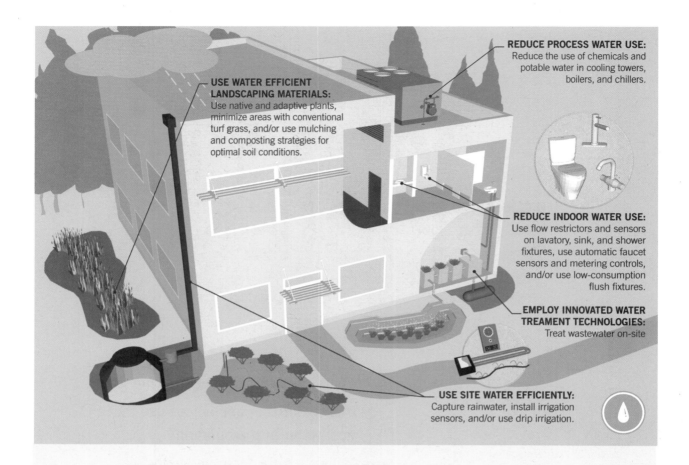

REDUCE PROCESS WATER USE:
Reduce the use of chemicals and potable water in cooling towers, boilers, and chillers.

USE WATER EFFICIENT LANDSCAPING MATERIALS:
Use native and adaptive plants, minimize areas with conventional turf grass, and/or use mulching and composting strategies for optimal soil conditions.

REDUCE INDOOR WATER USE:
Use flow restrictors and sensors on lavatory, sink, and shower fixtures, use automatic faucet sensors and metering controls, and/or use low-consumption flush fixtures.

EMPLOY INNOVATED WATER TREAMENT TECHNOLOGIES:
Treat wastewater on-site

USE SITE WATER EFFICIENTLY:
Capture rainwater, install irrigation sensors, and/or use drip irrigation.

WATER EFFICIENCY OVERVIEW

In this chapter, you will read a lot about water—something so vital to our daily lives yet often taken for granted. Between increasing demand and shrinking supply, our water resources are strained, threatening both human health and the environment. In short, the current trend in the demand for water is completely unsustainable, with many cities projecting serious water shortages within 10 years.

An easy way to organize your thinking about water efficiency is to consider it based on the three areas of use for a project.

- **Indoor water** — water that is used inside the building for toilet flushing, hand washing, and drinking.

- **Irrigation water** — water that is used outside the building to maintain landscaping.

- **Process water** — water that is used in building systems to heat and cool air to maintain building temperatures. This also includes water used for dishwashing or washing machines.

Environment: The demand for water drives a slew of operations that have negative environmental impacts. In order to keep a supply of water, we build dams, dig wells, and make withdrawals from our natural water bodies. The built environment also makes it increasingly difficult for water to naturally recharge the system.

Economy: Talk about resources that aren't priced according to their true value—water has to be one of the most underpriced commodities. While the economics of saving water today might not be that impressive, as supply dwindles, the benefits become more apparent. Another cost-saving measure is that when you use less water, you also use less energy to heat that water.

Community: There is no doubt that water is different than any other commodity; it is fundamental to our existence. Ensuring safe, clean water for future generations is an imperative that can't be ignored.

Indoor Water

As you've read, the primary way to improve the quality and availability of water is to use less of it. Simple, right?

Strategies

Actually, changing behavior regarding water use can be relatively simple. Efficiency measures, many of which go unnoticed by building occupants, can significantly and easily reduce the amount of water used in buildings.

Strategies

Waterless Urinal

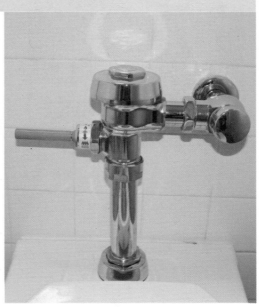

Dual Flush Toilet

Use efficient fixtures.

Nowadays, it is pretty easy to obtain quality fixtures that use significantly less water than their predecessors. Low-flow toilets, faucets, and showerheads all provide the same service and use far less water.

Use nonpotable water.

Sure, you can't drink the water that falls on your roof, but why not flush your toilet with it? Collecting or harvesting rainwater for use in toilets can take the strain off municipal potable water supplies.

Photo by Gary Crawford

Install meters.

It's hard to fix a problem if you don't know what the problem is. Meters identify where water is being used so you can tell if you have a leak or some other issue.

Outdoor Water

What about outside? Outdoor water uses, primarily landscaping, account for more than a quarter of all water consumption.

Strategies

Improved landscape design and maintenance can greatly reduce or even eliminate the need for potable water outdoors.

Photo by MDRP/Michael David Rose

Strategies

Choose locally adapted plants.

Plants that thrive naturally in an area do so for a reason — they have adapted to the soil and climate and therefore require little maintenance.

Photo by James Simmons

Use xeriscaping.

Xeriscaping is another term for landscaping design that focuses on using plants that are native to an area and therefore require little water. In addition, xeriscaping emphasizes soil improvements, mulching, and efficient irrigation to maximize any water that is used.

Select efficient irrigation technologies.

Sprinklers are great for running through on a hot summer day, but they are certainly not the most efficient way to water plants. Plants drink from their roots, therefore, supplying water directly to their roots means the plants get the bulk of the water instead of it evaporating into the air.

Photo by Jennifer L. Owens

Use nonpotable water.   Nonpotable water refers to water that isn't fit for humans to drink but is safe for watering plants. Capturing rainwater from the roof and using it as water for irrigation is one option for using the water that falls on the site.

Install submeters.   Again, it's hard to fix a problem when you can't find it. Submeters help identify leaks and let you know where your water is going.

Process Water

Process water refers to all the other water used in the building other than the water talked about above. This includes the water used in dishwashers, clothes washers, and ice machines. It also refers to water used for industrial purposes and in building systems such as boilers and chillers.

Strategies

At the building management level, there are numerous opportunities to reduce the amount of water being used for operational and industrial purposes.

Strategies

Use efficient fixtures.

Water-efficient appliances are easy to come by and are cost competitive, thanks to increased consumer demand.

Use nonpotable water.

Well-designed building systems don't need to use excessive amounts of potable water. Closed-loop systems keep water clean in a contaminant-free closed loop that stretches the use of the water.

Install submeters.

Process water systems can have leaks and maintenance issues. Metering helps quickly identify problem areas.

WE CATEGORY REVIEW

What are the goals of the Water Efficiency category?

What are some uses of nonpotable water?

What are some strategies for reducing water consumption?

LEARNING ACTIVITIES

THINK ABOUT IT

Which of the following water efficiency systems do you use or have you seen?

- Waterless urinals
- Dual-flush toilets
- Low-flow showerheads and faucets
- Drip irrigation systems
- Submeters
- Rainwater-harvesting system

If you haven't seen one of these systems, see if you can find one in your community.

GROUP ACTIVITY

Provide an image of a building using several water efficiency strategies. Show the image so that all participants can see it, or pass out a copy of it. Give everyone two minutes to list as many water efficiency strategies as they see. Share the results with everyone.

TRY IT OUT

Write on the board or a piece of paper the three types of water usage—indoor, outdoor, and process. Then place the items below in the correct category. Further identify which of these items could potentially use nonpotable water.

- Toilets
- Boilers
- Drinking fountains
- Garden hoses
- Dishwashers
- Bathroom faucets
- Cooling tower
- Drip irrigation system

PRACTICE QUESTIONS

1. What is the primary standard used to establish the baseline case for indoor water use?

a.) The Clean Water Act
b.) Energy Policy Act of 1992
c.) ASHRAE Standard 90.1
d.) National Environmental Policy Act

2. What is reduced when a project uses reclaimed water in its cooling towers?

a.) Potable water use
b.) Process water use
c.) Indoor plumbing water use
d.) Nonpotable water use
e.) Irrigation water use

3. How can potable water use for irrigation be reduced or eliminated (select two)?

a.) Install submeters
b.) Select locally adapted plants
c.) Increase the coverage of turf grass
d.) Use organic fertilizers
e.) Select noninvasive plants

4. Wastewater from toilets and urinals is known as ___?

a.) Brownwater
b.) Potable water
c.) Blackwater
d.) Graywater
e.) Drinking water

5. Nonpotable water is typically suitable for which of the following uses (select two)?

a.) Ice making
b.) Drinking
c.) Showers
d.) Toilet flushing
e.) Irrigation

6. What is the baseline water use for water closets?

a.) 0.8 gallons per flush
b.) 1.0 gallons per flush
c.) 1.6 gallons per flush
d.) 3.2 gallons per flush

7. Municipally supplied reclaimed water is considered ___?

a.) Free water
b.) Nonpotable water
c.) Process water
d.) Blackwater

8. Process water use can be reduced by ___? (select two)

a.) Installing submeters
b.) Installing low-flow showerheads
c.) Using high-efficiency irrigation technologies
d.) Using ENERGY STAR–certified clothes washers

Answer Key on Page 135

KEY TERMS TO KNOW:

Cover up the left side of the page and test yourself to see if you can summarize the definitions of the following key terms. Or better yet, make flash cards.

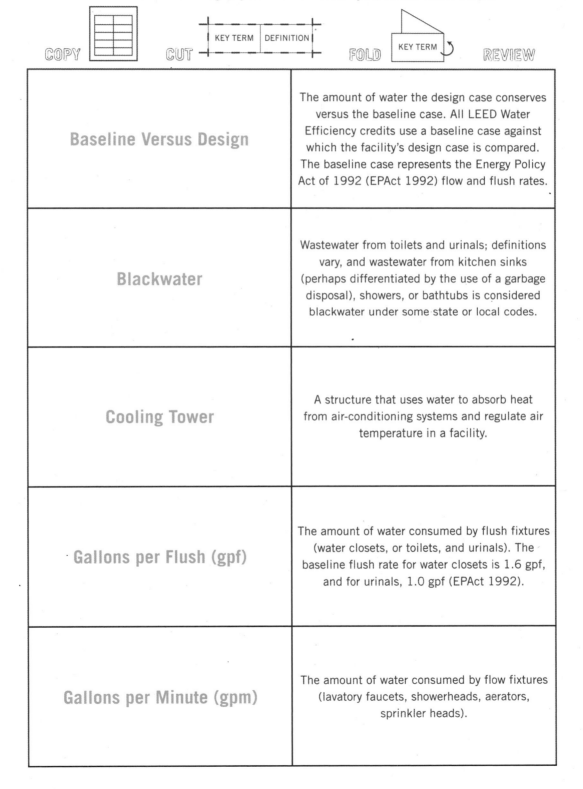

COPY   CUT   KEY TERM | DEFINITION   FOLD   KEY TERM   REVIEW

Baseline Versus Design	The amount of water the design case conserves versus the baseline case. All LEED Water Efficiency credits use a baseline case against which the facility's design case is compared. The baseline case represents the Energy Policy Act of 1992 (EPAct 1992) flow and flush rates.
Blackwater	Wastewater from toilets and urinals; definitions vary, and wastewater from kitchen sinks (perhaps differentiated by the use of a garbage disposal), showers, or bathtubs is considered blackwater under some state or local codes.
Cooling Tower	A structure that uses water to absorb heat from air-conditioning systems and regulate air temperature in a facility.
Gallons per Flush (gpf)	The amount of water consumed by flush fixtures (water closets, or toilets, and urinals). The baseline flush rate for water closets is 1.6 gpf, and for urinals, 1.0 gpf (EPAct 1992).
Gallons per Minute (gpm)	The amount of water consumed by flow fixtures (lavatory faucets, showerheads, aerators, sprinkler heads).

Graywater	Domestic wastewater composed of wash water from kitchen, bathroom, and laundry sinks, tubs, and washers. (EPA)
Irrigation Efficiency	The percentage of water delivered by irrigation equipment that is actually used for irrigation and does not evaporate, blow away, or fall on hardscape. For example, overhead spray sprinklers have lower irrigation efficiencies (65%) than drip systems (90%).
Harvested Rainwater	Precipitation captured and used for indoor needs, irrigation, or both.
Potable Water	Water that meets or exceeds the EPA's drinking water quality standards and is approved for human consumption by the state or local authorities having jurisdiction; it may be supplied from wells or municipal water systems.
Wastewater	The spent or used water from a home, community, farm, or industry that contains dissolved or suspended matter.

WHAT ABOUT ENERGY AND ATMOSPHERE?

- Do you know where your electricity comes from?

- How much energy do buildings really use?

- Why should buildings generate all of their energy on-site?

- Why do some developers build energy-inefficient buildings?

- Can a safe, comfortable building sell electricity back to its community?

- Should a home have air conditioning? What could be gained? What would be lost?

Read pages 64–70 of the Green Building and LEED Core Concepts Guide, Second Edition, or pages 43-51 in the First Edition.

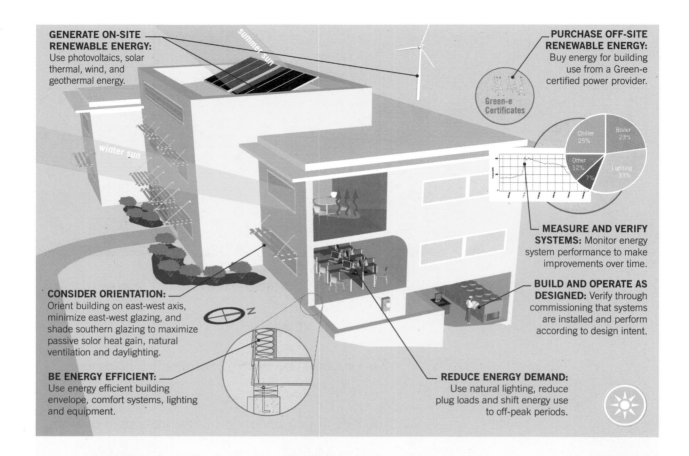

GENERATE ON-SITE RENEWABLE ENERGY: Use photovoltaics, solar thermal, wind, and geothermal energy.

summer sun

PURCHASE OFF-SITE RENEWABLE ENERGY: Buy energy for building use from a Green-e certified power provider.

Green-e Certificates

winter sun

Chiller 25%
Boiler 23%
Other 12%
7%
Lighting 33%

MEASURE AND VERIFY SYSTEMS: Monitor energy system performance to make improvements over time.

CONSIDER ORIENTATION: Orient building on east-west axis, minimize east-west glazing, and shade southern glazing to maximize passive solar heat gain, natural ventilation and daylighting.

BUILD AND OPERATE AS DESIGNED: Verify through commissioning that systems are installed and perform according to design intent.

BE ENERGY EFFICIENT: Use energy efficient building envelope, comfort systems, lighting and equipment.

REDUCE ENERGY DEMAND: Use natural lighting, reduce plug loads and shift energy use to off-peak periods.

ENERGY AND ATMOSPHERE OVERVIEW

In this chapter, you will learn about the importance of energy performance in buildings and reviewed strategies to reduce the environmental impacts associated with energy use. Additionally, the environmental trade-offs of refrigerants have been explored. On the one hand, they have remarkable physical properties that enable efficient cooling, but, on the other hand, they directly contribute to global warming and ozone depletion.

Strategies to reduce the impacts of energy use focus on four, interconnected elements:

- **Energy demand;**
- **Energy efficiency;**
- **Renewable energy; and**
- **Ongoing energy performance.**

Energy Use

Many of the primary methods of energy extraction have negative environmental effects. Intentionally using energy only where it is needed, followed by ensuring that efficient systems are used to meet these needs, is the most effective way to reduce the environmental impacts of energy use. The use of less harmful forms of energy, such as energy generated on-site from renewable sources or Green-e–certified electricity, can further reduce the negative effects of energy. Finally, it is important to ensure sustained performance over time through implementation of building systems commissioning and metering of the building energy systems.

Environment: Energy production causes emissions that contribute to climate change, smog, and reduced air quality on a global scale. With the enormous amount of energy that is used within buildings comes an opportunity to reduce these environmental impacts.

Between 1990 and 2006, annual greenhouse gas emissions from U.S. energy use increased by 16%. Buildings consume approximately 39% of the energy and 74% of the electricity produced annually by the United States, according to the Department of Energy.

Refrigerants are a necessary component in the refrigeration cycle employed by most air-conditioning and refrigeration equipment. Many refrigerants, however, are harmful to the environment and human health.

Economy: Energy costs represent a significant percentage of buildings' operating expenses. Reductions in energy use, and therefore energy costs, can benefit the economy, allowing businesses to expand and/or invest in further building improvements.

Community: Efficient buildings within a city or region can serve as a catalyst for community engagement and foster a sense of civic pride. A community that, as a whole, significantly reduces its energy use can benefit from both the direct savings (from lower energy use), thereby increasing the availability of funds for local investment, as well as the potential savings due to the lessened need for additional energy-generation infrastructure.

Figure 1 from the LEED Reference Guide for Green Building Operations & Maintenance, 2009. Page 137. Example of Energy Use Breakdown.

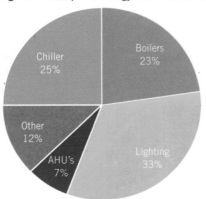

Other factors also affect energy usage, including other aspects of building performance that have a direct impact on energy use. Some examples follow:

- Water use can directly affect energy use because it typically takes gas or electricity to heat the water and to move the water to and from the building.

- Interior color schemes can significantly influence energy use. Light-color surfaces reflect more light than dark surfaces and can reduce the total amount of lighting power necessary to maintain the intended light levels.

- Increased ventilation strategies, while helpful for increasing the overall quality of indoor air, typically require a corresponding increase in HVAC energy use.

- Lighting to meet the needs and preferences of building occupants should be a main consideration of project teams to ensure that the use of the space requires minimal energy. Enabling occupants to turn off their lights when not in use can net significant lighting energy savings.

Conventional incandescent lamps use approximately four times the energy to create the same light output as standard compact fluorescent lamps. The wasted energy is converted into heat and, in many buildings, creates an additional burden on the air-conditioning system.

Energy Demand

Addressing energy demand is the first step toward reducing the negative environmental consequences of energy use. Strategies to reduce energy demand are as follows:

Strategies

Establish design and energy goals.

You can't achieve a goal if you don't take the time to establish one. Work with all the members of the design team early on to figure out what makes sense for your project. For example, would it make sense to install a radiant floor for heating rather than a forced-air system? What would the energy bills be over the course of a year to operate the building? What design decisions could lower that number?

Size the building appropriately.

Early on in the project, the project team should critically examine the needs of the future occupants and design the building to meet, but not exceed, those needs.

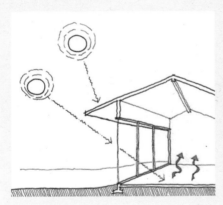

Use free energy.

There is absolutely nothing more efficient than taking advantage of what Mother Nature offers to us free of charge. When determining how the building will be heated, cooled, lit, and ventilated, explore natural sources first. Many building sites have the potential to satisfy the majority of human needs through passive systems such as daylighting, natural ventilation, and using the building mass as a thermal storage system.

© 2007 The Green Studio Handbook: Environmental Strategies for Schematic Design, Alison G. Kwok, Walter T. Grondzik; used with permission.

Insulate.

As simple as it sounds, insulation helps keep inside the building the air that owners have paid to heat or cool. High-performance building envelopes insulate efficiently, allowing for significant reductions in energy use and, potentially, for downsizing of heating and cooling systems.

Monitor consumption.

If you aren't paying attention, you really won't know whether people are using the building's systems in the way they should. Building occupants can be encouraged to respond to alerts and notifications that are initiated when energy demand exceeds certain levels or continues for specified durations.

Energy Efficiency

Energy efficiency builds on the strategies used to reduce energy demand. Once energy needs have been identified by exploring energy demand reductions, high-efficiency systems can be used to further drive down energy use.

Photo by Kalpana Kuttaiah

Strategies

Identify passive design opportunities.

The idea of passive design goes back to using the natural attributes of a site. Buildings that provide access to natural daylight, and are supported by active controls, can achieve significant lighting energy savings simply by using the sun instead of artificial lighting.

Photo by K.C. Kratt

Strategies

Address the building envelope.

The extent to which envelope systems (such as glazing and insulation) are appropriate for a specific project is determined largely by the climatic conditions where the project is located. Projects in extreme climates (seasonally very hot or very cold) benefit more from higher levels of insulation.

Photo by ©Christian Richters

Install high-performance mechanical systems.

High-performing systems often carry a cost premium but use less energy than conventional systems. Investments in premium systems often pay for themselves many times over during the equipment's lifetime.

Specify high-efficiency appliances.

Look for the star! ENERGY STAR equipment uses less energy than conventional appliances. The ENERGY STAR label is an easily recognizable indicator of efficiency for eligible equipment types, such as computers, monitors, and refrigerators.

Photo by Jonathan Leys

Use energy simulation models.

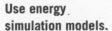

Building performance can be predicted before the building is even constructed, and design options can be evaluated to determine which one offers the greatest energy benefit. Energy simulation or energy modeling allows the team to look at the different energy efficiency measures to see how they work together and which offer the greatest benefit.

Strategies

Use high-efficiency infrastructure.

Infrastructure, such as street lighting and traffic control devices, creates opportunities for significant savings due to its long hours of use.

Photo courtesy of Russell T. Tutt Science Center

Use thermal energy storage.

Because of the day/night cycle and corresponding temperature fluctuations, it is possible to capture heat during the day for use at night, and reject heat at night to provide cooling during the day.

Photo by Colin Jewall and Enrico Dagostini

Capture efficiencies of scale.

Buildings can be heated or cooled by a district system (a system that provides thermal conditioning to multiple buildings). These systems represent important opportunities of scale due to their large size.

Renewable Energy

Generating power from renewable sources (such as solar, wind, and biomass) avoids air and water pollution and other harmful environmental effects associated with the production and consumption of traditional fuels (such as coal, nuclear power, oil, and natural gas). By using renewable energy sources wherever possible, consumption of limited fossil fuels and production of emissions are reduced.

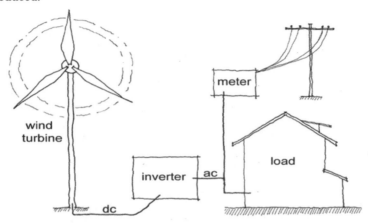

© 2007 The Green Studio Handbook: Environmental Strategies for Schematic Design, Alison G. Kwok, Walter T. Grondzik; used with permission.

Strategies

Generate on-site renewable energy.

Projects with access to direct sunlight, especially those in lower latitudes, can generate significant electricity using photovoltaic panels and reduce their energy use for hot water by using solar hot-water heaters.

Purchase off-site renewable energy.

Projects wishing to use lower-impact energy sources can purchase green power from their utility. Additionally, projects can purchase renewable energy certificates from local or national providers, regardless of their utility's direct offerings.

Ongoing Energy Performance

The standard increment of electricity is the kilowatt-hour (kWh). This represents the amount of electricity that would be put out by a 1 kW source in one hour.

Even the most carefully designed buildings, using an integrative design approach and implementing the best-performing systems, can underperform due to any number of items that can't be foreseen on the drafting table. Therefore, continued attention to energy performance is critical during construction and throughout the building's operation.

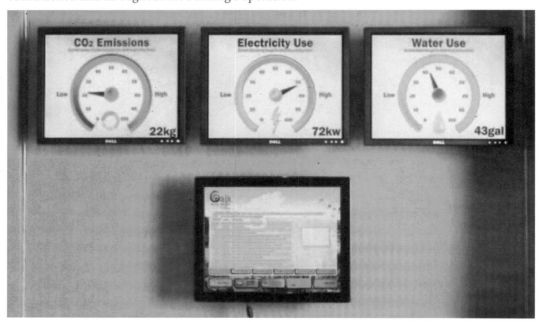

Photo by Wen Chang

Strategies

Adhere to owner's project requirements.

The owner's project requirements (OPR) are determined at the beginning of design and establish the building's functional requirements. During design, and throughout construction and operation, the OPR should be used as a benchmark for achievement when evaluating design options and functionally testing the systems once constructed.

Photo by Brad Feinknopf

Strategies

Provide staff training.

Informed building occupants are better able to use the systems as they are intended. Staff training resources ensure that building operators understand how to maintain and operate complex systems.

Photo by Eric Laignel

Conduct preventive maintenance.

Building systems that operate properly are able to fulfill their design function effectively and efficiently. Periodic maintenance helps ensure that systems stay in peak operating condition.

Create incentives for occupants and tenants.

Motivated building users can be an ally in reducing energy use. The use of incentives and the direct transfer of energy costs can help encourage occupants to turn off lights and equipment when they are not in use.

EA CATEGORY REVIEW

How can buildings reduce the negative environmental consequences associated with energy use?

What are some strategies to reduce energy demand?

How can buildings reduce the impact of energy used?

What are some LEED credits from other categories that are synergistic with energy use?

LEARNING ACTIVITIES

THINK ABOUT IT

Consider buildings in your community (these may be the same buildings you identified for earlier exercises). What energy efficiency features do they currently have? What features would be possibilities for them?

FEATURE	INSTALLED NOW/FUTURE/POSSIBILITY
Passive design	
Insulation/weatherization	
High-performance windows	
High-performance mechanical systems	
High-efficiency appliances	

GROUP ACTIVITY

Explore renewable opportunities. Consider a specific site within your community. This can be your home, school, office, or other location where the group has specific knowledge of its surrounding environment. Identify which renewable energy systems would work best given the site's natural resources and microclimate. Determine what possible mix of renewable energy systems could be integrated to provide a significant portion of the building's energy. Discuss the following questions:

- Based on the availability of natural resources (wind, sun, water, and so on), what systems are most appropriate for the site? What systems don't make sense?

- Is it possible to meet the building's energy needs entirely with on-site systems? Why or why not?

INVESTIGATE

List all energy uses at your house. Work in small groups to review each participant's list and identify additional uses. Add up all the energy uses at your house. List all assumptions. Return to the full group to discuss the assumptions and reach a consensus. Each participant should recalculate his or her energy use based on the group assumptions. Discuss the results.

SITE VISIT

Visit the Energy Information Administration Commercial Buildings Energy Consumption Survey at http://www.eia.doe.gov/emeu/cbecs/ and determine the per-square-foot energy intensity for the sum of major fuel consumption in 2003 (Table C3A) for a building in the education category.

PRACTICE QUESTIONS

1. **What is the first step a project team should consider when trying to save energy?**

   a.) Use on-site renewable energy.
   b.) Reduce energy demand.
   c.) Adopt energy efficiency measures.
   d.) Install submetering equipment.

2. **What strategy is being used for a project design that orients windows to allow the building to be warm in the winter, stay cool in the summer, and capture daylight?**

   a.) High-performance mechanical systems
   b.) Efficiencies of scale
   c.) Energy simulation modeling
   d.) Passive design concepts

3. **To what do renewable energy certificates (RECs) refer?**

   a.) On-site photovoltaic systems
   b.) Off-site renewable energy purchases
   c.) Market-generated wind power
   d.) Stock in utility providers

4. **One of the most cost-effective ways to ensure optimal ongoing energy performance is to ___?**

   a.) Commission building systems
   b.) Upgrade mechanical systems
   c.) Install renewable energy systems
   d.) Maintain trees and landscaping features

5. **Refrigerants are necessary as part of the refrigeration cycle often used to cool buildings. However, the benefits of their use should be considered against their potential for adverse environmental impacts. These direct impacts are quantified by which metrics (select two)?**

   a.) Global warming potential
   b.) Energy use potential
   c.) Ozone depletion potential
   d.) Soil contamination potential
   e.) Air pollution potential

6. **A project team is selecting the HVAC system for the tenant space. Which primary factors should be considered to minimize the environmental impact of the system (select two)?**

   a.) First cost
   b.) Expected life
   c.) Energy performance
   d.) Lead time

7. **Energy use associated with office equipment, kitchen cooking, and escalators is known as ___?**

   a.) Regulated energy
   b.) Exempt energy
   c.) Process energy
   d.) Secondary energy

8. **Which are considered renewable energy sources (select two)?**

   a.) Nuclear
   b.) Solar
   c.) Wave
   d.) Natural gas

Answer Key on Page 136

KEY TERMS TO KNOW:

Cover up the left side of the page and test yourself to see if you can summarize the definitions of the following key terms. Or better yet, make flash cards.

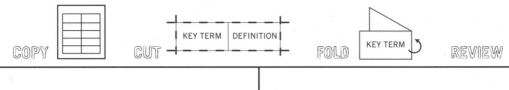

COPY   CUT   FOLD   REVIEW

British Thermal Unit (Btu)	The amount of heat required to raise the temperature of one pound of liquid water from 60° to 61° Fahrenheit. This standard measure of energy is used to describe the energy content of fuels and compare energy use.
Building Envelope	The exterior surface of a building—the walls, windows, roof, and floor; also referred to as the building shell.
Chiller	A device that removes heat from a liquid, typically as part of a refrigeration system used to cool and dehumidify buildings.
Compact Fluorescent Lamp (CFL)	A small fluorescent lamp, used as a more efficient alternative to incandescent lighting; also called a PL, twin-tube, or biax lamp. (EPA)
Energy-Efficient Products and Systems	Building components and appliances that use less energy to perform as well as or better than standard products.

Energy or Greenhouse Gas Emissions per Capita	A community's total greenhouse gas emissions divided by the total number of residents.
Energy Management System	A control system capable of monitoring environmental and system loads and adjusting HVAC operations accordingly in order to conserve energy while maintaining comfort. (EPA)
ENERGY STAR® Rating	A measure of a building's energy performance compared with that of similar buildings, as determined by the ENERGY STAR Portfolio Manager. A score of 50 represents average building performance.
Energy Use Intensity	Energy consumption divided by the number of square feet in a building, often expressed as British thermal units (Btus) per square foot or as kilowatt-hours of electricity per square foot per year (kWh/sf/yr).
Fossil Fuel	Energy derived from ancient organic remains, such as peat, coal, crude oil, and natural gas. (EPA)
HVAC Systems	Equipment, distribution systems, and terminals that provide the processes of heating, ventilating, or air-conditioning. (ASHRAE Standard 90.1–2007)

Lighting Power Density	The installed lighting power per unit area.
Measures of Energy Use	Typical primary measures of energy consumption associated with buildings include kilowatt-hours of electricity, therms of natural gas, and gallons of liquid fuel.
Nonrenewable	Not capable of being replaced; permanently depleted once used. Examples of nonrenewable energy sources are oil and natural gas; nonrenewable natural resources include metallic ores.
Performance Relative to Benchmark	A comparison of a building system's performance with a standard, such as ENERGY STAR Portfolio Manager.
Performance Relative to Code	A comparison of a building system's performance with a baseline that is equivalent to minimal compliance with an applicable energy code, such as ASHRAE Standard 90 or California's Title 24.
Photovoltaic (PV) Energy	Electricity from photovoltaic cells that convert the energy in sunlight into electricity.

Renewable Energy	Resources that are not depleted by use. Examples include energy from the sun, wind, and small (low-impact) hydropower, plus geothermal energy and wave and tidal systems. Ways to capture energy from the sun include photovoltaic, solar thermal, and bioenergy systems based on wood waste, agricultural crops or residue, animal and other organic waste, or landfill gas.
Renewable Energy Certificate (REC)	A tradable commodity representing proof that a unit of electricity was generated from a renewable energy resource. RECs are sold separately from the electricity itself and thus allow the purchase of green power by a user of conventionally generated electricity.

WHAT ABOUT
MATERIALS AND
RESOURCES?

- How do you know if a building material is safe?

- Where do the majority of building materials come from?

- What are the most sustainable building materials? What criteria do you use to evaluate them?

- What happens to leftover construction materials?

- Which materials take the longest to break down at the landfill?

Read pages 71-76 of the Green Building and LEED Core Concepts Guide, Second Edition, or pages 53-57 in the First Edition.

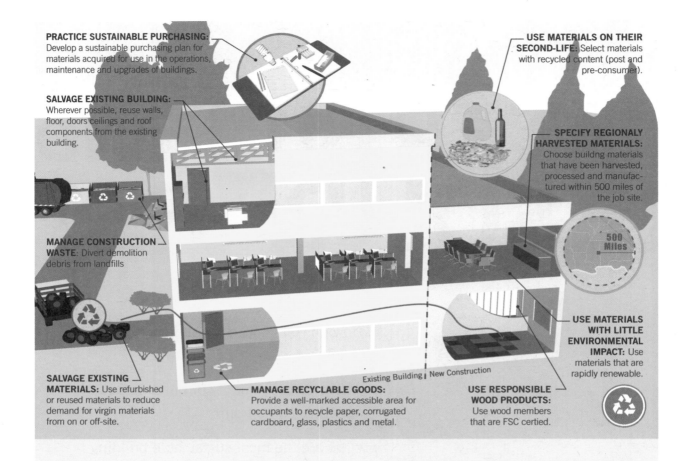

PRACTICE SUSTAINABLE PURCHASING: Develop a sustainable purchasing plan for materials acquired for use in the operations, maintenance and upgrades of buildings.

SALVAGE EXISTING BUILDING: Wherever possible, reuse walls, floor, doors ceilings and roof components from the existing building.

MANAGE CONSTRUCTION WASTE: Divert demolition debris from landfills

SALVAGE EXISTING MATERIALS: Use refurbished or reused materials to reduce demand for virgin materials from on or off-site.

MANAGE RECYCLABLE GOODS: Provide a well-marked accessible area for occupants to recycle paper, corrugated cardboard, glass, plastics and metal.

USE MATERIALS ON THEIR SECOND-LIFE: Select materials with recycled content (post and pre-consumer).

SPECIFY REGIONALY HARVESTED MATERIALS: Choose buildng materials that have been harvested, processed and manufactured within 500 miles of the job site.

500 Miles

USE MATERIALS WITH LITTLE ENVIRONMENTAL IMPACT: Use materials that are rapidly renewable.

Existing Building | New Construction

USE RESPONSIBLE WOOD PRODUCTS: Use wood members that are FSC certied.

MATERIALS AND RESOURCES OVERVIEW

In this chapter, you will be challenged to consider where materials come from and where they go. From their extraction on through their sale, use, and disposal, materials have extensive impacts on our lives and the environment.

Green building focuses on two primary categories when considering materials and resources:

- **Selection of sustainable materials (life-cycle impacts); and**
- **Waste management.**

Material Selection—Considering Life-Cycle Impacts

Obviously the development of buildings requires the use of materials ... a lot of materials! In fact, 40% of raw stone, gravel, and sand and 25% of all virgin wood that is extracted is used for the construction of buildings.

Environment: Given the sheer amount of resources that go into buildings, it is easy to see the environmental consequences of their use, including air and water pollution, destruction of native habitats, and depletion of natural resources.

Economy: Many of the readily available materials used in the construction industry are extracted or manufactured in a way that historically has focused solely on economic profit. In this model, the environment and people are completely left out of the picture.

Community: The extraction, manufacturing, application, and disposal of many of today's materials pose serious health risks to the human population, many of which are not clearly understood.

Strategies

So what exactly does it mean when you hear the term "life cycle" in regard to material selection? Basically, it means you are considering all the environmental impacts associated with a material from the beginning (i.e. its extraction) through its processing and use in manufacturing, its transportation, and its end use, all the way to its eventual disposal. A life-cycle assessment, or LCA, is a "cradle-to-grave" approach for assessing materials. Cradle-to-grave begins with the gathering of raw materials from the earth to create the product and ends at the point when all materials

Photo courtesy of the John Buck Company

are returned to the earth. While you may not be able to complete a formal LCA on all the products and services you use on a project, the framework itself can be extremely useful when weighing various options.

Strategies

Develop a construction purchasing policy.

There are a lot of people involved in the design and construction of a project, and not all of them necessarily understand your goals. Put all these goals into a construction purchasing policy so that there is no question about what products you'd like to see selected for the project.

Specify green materials.

Specifications are the written documents that an architect issues along with drawings to describe the quality of materials to be used on a project. It is in the specifications that you can let the contractor know to buy wood from the region or not to use PVC, for example. The specifications are a legally binding document, so it is the most solid way to ensure that green materials make it into the project.

Photo by Jonathan Leys

Specify green interiors.

The inside of a building is one of the most critical areas for human health to specify green. As noted above, you can use the specifications to make sure only low- or no-VOC paints are used or bamboo floors, for example, are installed.

Develop a sustainable purchasing policy.

The word "policy" is enough to scare a lot of people off, but really a purchasing policy can be a simple checklist that documents the types of products you want to be used in your project. It ensures that everyone from the top down understands what your goals are and how these translate into purchases so that the paper with 100% recycled content makes it to the copier instead of virgin paper.

Specify green electronic equipment.

There are already countless choices when it comes to electronic equipment, but thinking about energy use and material selection is something you can't forget about. Find electronics and accessories that use less energy, are made with recyclable and recycled materials, and are easier to upgrade, fix, and recycle.

Waste Management

The funny thing about throwing something away is that there is no such thing as "away." The three necessary components for decomposition—sunlight, moisture, and oxygen—are hard to come by in a landfill. Although we're not sure, since we haven't been paying attention long enough, scientists estimate that plastic could take 500 years or longer to decompose in a landfill.

And where is all this garbage coming from? Construction and demolition of buildings account for about 40% of the total waste stream in the United States.

Environment: Current construction and disposal practices introduce highly hazardous toxic materials into the soil, air, and water when such materials are disposed of in landfills. Landfills fail and leak contaminants into ground and surface water.

Economy: The cost of building and maintaining landfills is astronomical. Because landfills have the potential to substantially harm the environment, they are highly regulated and therefore costly. Even after a landfill is "closed," it must be monitored and tested for decades to determine its impact on local air, soil, and water resources.

Community: Landfill gas emissions and contaminated groundwater sources have the potential for devastating impacts on human health.

Strategies

OK, so you've heard enough. But what's the answer? Well, you know the three R's, right? Reduce, reuse, recycle. But did you know that they're in that order for a reason? It's a hierarchy, and it's best to start at the top with reduce. That's because reducing the amount of materials minimizes the environmental impacts throughout each phase of the materials life cycle. Next comes reusing, which means that instead of building a new building or buying new cabinets, you look at reusing existing products that may just need to be fixed up. Last of the R's is recycling, which means taking waste to recycling facilities that break down the materials and turn them back into "new" materials.

Good waste management can reduce the amount of waste and toxins hauled to and disposed of in landfills or incineration facilities. With that in mind, consider the strategies below, each of which fits into either the reduce, reuse, or recycle category.

Strategies

Size the building appropriately.

Bigger isn't always better. Buildings that are smartly designed meet the needs of their occupants without wasting space. Using less means you inherently waste less.

Reuse existing buildings or portions of existing buildings.

Given the number of existing buildings in this country, we need to get more creative about repurposing them. Consider an old manufacturing facility that has sat unused for years. It could have great "bones," and some creative design decisions could transform it into a unique office space.

Photo by Melva P. Calder

Reuse building materials.

Certain salvaged materials simply cannot be duplicated in today's manufacturing system. Looking for architectural gems or repurposing a product such as old lockers is a great way to keep materials out of the landfill and give a space character.

Photo by Dale Photographic

Develop a construction waste management policy.

During construction, it is important to write down your goals so that you can hold everyone on-site accountable for the goals and processes you put in place. A construction waste policy should outline the three R's and spell out for all the trades how they can contribute to reducing the amount of waste headed to the landfill.

Photo by Jim Gallop/Gallop Studio

Consider new technology, design, and construction decisions.

Every day, designers and engineers are coming up with new ideas for addressing the myriad of issues associated with waste. Such strategies might include a stained concrete floor (design strategy), advanced wood framing (construction strategy), or carpet tiles (new technology strategy).

Compost.

One person's waste is another person's meal ... er, another plant's meal. Food and landscaping waste can easily be transformed into mulch for garden beds.

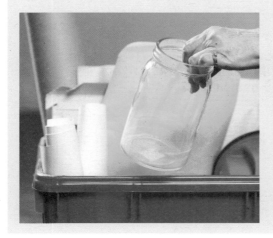

Encourage recycling.

It's pretty basic, but make sure it's as easy for building occupants to recycle as it is for them to throw something away. This means not only dedicating spaces for recycling but educating occupants on what goes where.

MR CATEGORY REVIEW

List some strategies for reducing the environmental impact associated with materials use.

What elements contribute to the life-cycle impacts of a material?

When considering materials, what environmental characteristics are important to prioritize, and what should be avoided?

LEARNING ACTIVITIES

WATCH IT

Watch "The Story of Stuff" online for free at:

http://www.storyofstuff.com/index.html.

"The Story of Stuff" is a 20-minute, fast-paced, fact-filled look at the underside of our production and consumption patterns. It exposes the connections between a huge number of environmental and social issues and calls us together to create a more sustainable and just world. It will teach you something, it will make you laugh, and it just may change the way you look at all the stuff in your life forever.

TRY IT OUT

Try out the life-cycle impact software listed at the end of the Material and Resources Chapter in the Green Building and LEED Core Concepts Guide.

- BEES 4.0
- Construction carbon calculator
- ATHENA EcoCalculator for Assemblies
- EPEAT

How can you apply any of these to your projects?

THINK ABOUT IT

Select an ongoing consumable product (such as lamps, toilet paper, or hand sanitizer) and identify at least four criteria that should be considered when developing an ongoing purchasing policy for this product.

INVESTIGATE

Take a look around the room and identify some basic material components such as the flooring, paint, and furniture. Are there greener options than what is currently there?

PRACTICE QUESTIONS

1. **How many years can an agricultural product grow or be raised to be considered as rapidly renewable by LEED?**

   a.) 5
   b.) 10
   c.) 15
   d.) 20

2. **What is the first step in a successful waste management policy?**

   a.) Recycle all possible construction materials.
   b.) Reuse existing materials.
   c.) Reduce the total quantity of waste.
   d.) Specify recyclable materials.
   e.) Determine the embodied energy of the product.

3. **A project generates 100 tons of waste throughout construction. 50 tons are collected on-site and sent to a sorting facility with a facility-wide diversion rate of 60%. 40 tons are separated on-site and sent directly to recyclers. The final 10 tons are incinerated off-site. What percentage of waste does LEED consider to have been recycled?**

   a.) 40%
   b.) 50%
   c.) 70%
   d.) 80%

4. **LEED defines regional materials as originating within ___?**

   a.) 50 miles of the project site
   b.) 250 miles of the project site
   c.) 500 miles of the project site
   d.) 750 miles of the project site

5. **According to the EPA, what percentage of solid waste is currently recycled in the United States?**

   a.) 11%
   b.) 32%
   c.) 51%
   d.) 72%

6. **A building material that is made from recycled soda bottles contains ___?**

   a.) Post-industrial recycled content
   b.) Post-consumer recycled content
   c.) Pre-consumer recycled content
   d.) Pre-fabricated recycled content

7. **Ongoing consumables are consumed during which of the following?**

   a.) Design only
   b.) Construction only
   c.) Occupancy only
   d.) Both construction and operation
   e.) Construction and at the building's end of life

8. **Which environmentally preferable attributes of materials does LEED recognize (select three)?**

   a.) Are harvested and manufactured regionally
   b.) Contain recycled content
   c.) Are sourced from developing countries
   d.) Are salvaged
   e.) Offer future customization options

Answer Key on Page 136

KEY TERMS TO KNOW:

Cover up the left side of the page and test yourself to see if you can summarize the definitions of the following key terms. Or better yet, make flash cards.

COPY CUT | KEY TERM | DEFINITION | FOLD KEY TERM REVIEW

By-Product	A material, other than the principal product, generated as a consequence of an industrial process or as a breakdown product in a living system. (EPA)
Certified Wood	Wood that has been issued a certificate from an independent organization with developed standards of good forest management. This certificate verifies that wood products come from responsibly managed forests.
Construction and Demolition Debris	Waste and recyclables generated from construction and from the renovation, demolition, or deconstruction of existing structures. It does not include land-clearing debris, such as soil, vegetation, and rocks.
Construction Waste Management Plan	A plan that diverts construction debris from landfills through recycling, salvaging, and reuse.
Regional/Locally Sourced Materials	Also known as regional materials, the amount of a building's materials that are extracted, processed, and manufactured close to a project site, expressed as a percentage of the total materials cost. For LEED, regional materials originate within 500 miles of the project site.

Recycled Content	The percentage of material in a product that is recycled from the manufacturing waste stream (preconsumer waste) or the consumer waste stream (postconsumer waste) and used to make new materials. For LEED, recycled content is typically expressed as a percentage of the total material volume or weight.
Post-Consumer Recycled Content	The percentage of material in a product that was consumer waste. The recycled material was generated by household, commercial, industrial, or institutional end users and can no longer be used for its intended purpose. This includes returns of materials from the distribution chain. Examples include construction and demolition debris, materials collected through recycling programs, discarded products (such as furniture, cabinetry, and decking), and landscaping waste (such as leaves, grass clippings, and tree trimmings).
Pre-Consumer Recycled Content	The percentage of material in a product that was recycled from manufacturing waste. Preconsumer content was formerly known as postindustrial content. Examples include planer shavings, sawdust, bagasse, walnut shells, culls, trimmed materials, overissue publications, and obsolete inventories. Excluded are rework, regrind, or scrap materials capable of being reclaimed within the same process that generated them.
Rapidly Renewable Materials and Products	The amount of a building's agricultural products (fiber or animal) that are quickly grown or raised and can be harvested in a sustainable fashion, expressed as a percentage of the total materials cost. For LEED, rapidly renewable materials take 10 years or less to grow or raise.
Reuse	The amount of building materials returned to active use (in the same or a related capacity as their original use), expressed as a percentage of the total materials cost of a building. The salvaged materials are incorporated into the new building, thereby extending the lifetime of materials that would otherwise be discarded.
Salvaged Material	Construction items recovered from existing buildings or construction sites and reused. Common salvaged materials include structural beams and posts, flooring, doors, cabinetry, brick, and decorative items.

Sustainable Forestry	Management of forest resources to meet the long-term forest product needs of humans while maintaining the biodiversity of forested landscapes. The primary goal is to restore, enhance, and sustain a full range of forest values, including economic, social, and ecological considerations.
Sustained-Yield Forestry	Management of a forest to produce in perpetuity a high-level annual or regular periodic output, through a balance between increment and cutting. (Society of American Foresters)
Waste Diversion	The amount of waste disposed of other than through incineration or in landfills, expressed in tons. Examples of waste diversion include reuse and recycling.

WHAT ABOUT INDOOR ENVIRONMENTAL QUALITY?

- What is your favorite place to be in your house? Why?

- How much time, on average, do you spend inside?

- Why is it beneficial for occupants to have individual lighting controls?

- What is the point of access to views of the outdoors?

- Why do buildings use entryway systems such as walk-off mats?

- Is it worthwhile to provide more ventilation than required by code?

- Why are acoustics of special concern to school classrooms?

- What's so great about daylight?

Read pages 77-82 of the Green Building and LEED Core Concepts Guide, Second Edition, or pages 59-63 in the First Edition.

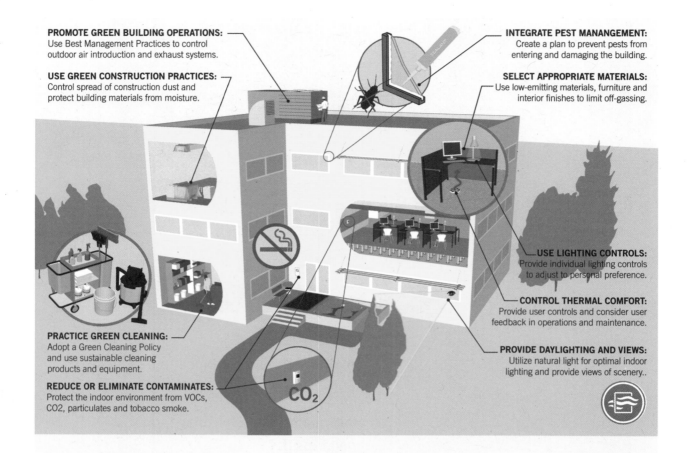

PROMOTE GREEN BUILDING OPERATIONS: Use Best Management Practices to control outdoor air introduction and exhaust systems.

USE GREEN CONSTRUCTION PRACTICES: Control spread of construction dust and protect building materials from moisture.

INTEGRATE PEST MANANGEMENT: Create a plan to prevent pests from entering and damaging the building.

SELECT APPROPRIATE MATERIALS: Use low-emitting materials, furniture and interior finishes to limit off-gassing.

USE LIGHTING CONTROLS: Provide individual lighting controls to adjust to personal preference.

CONTROL THERMAL COMFORT: Provide user controls and consider user feedback in operations and maintenance.

PROVIDE DAYLIGHTING AND VIEWS: Utilize natural light for optimal indoor lighting and provide views of scenery..

PRACTICE GREEN CLEANING: Adopt a Green Cleaning Policy and use sustainable cleaning products and equipment.

REDUCE OR ELIMINATE CONTAMINATES: Protect the indoor environment from VOCs, CO2, particulates and tobacco smoke.

INDOOR ENVIRONMENTAL QUALITY OVERVIEW

In this chapter, you will learn about the importance of maintaining high-quality indoor environments by controlling pollutants, introducing daylight and views, and putting some control into the hands of the occupants. Additionally, acoustics and thermal comfort were noted as critical to overall occupant satisfaction.

To summarize, the critical components of the Indoor Environmental Quality section are as follows:

- **Indoor air quality;**
- **Thermal comfort;**
- **Lighting; and**
- **Acoustics.**

Indoor Air Quality

Most of the air that building occupants breathe is delivered through the building's ventilation system (naturally or mechanically), and the quality of this air can be compromised by contaminants within the building. Indoor air quality is known to affect human health and can directly influence productivity and quality of life.

Photo courtesy of Exelon

Environment: Improving indoor air quality specifically benefits the places where Americans spend most of their time: indoors. Providing ample, but not excessive, ventilation strikes the right balance between energy use and human health. Too much air is wasteful in terms of energy use, and too little air can contribute to poor-quality indoor environments.

Economy: The potential annual savings and productivity gains from improved indoor environmental quality in the United States are estimated as follows[1]:

- $6 billion to $14 billion from reduced respiratory disease;

- $1 billion to $4 billion from reduced allergies and asthma;

- $10 billion to $30 billion from reduced sick-building–syndrome symptoms; and

- $20 billion to $160 billion from direct improvements in worker performance unrelated to health.

[1] IAQ Fact Sheet (Environmental Health, Safety and Quality Management Services for Business and Industry, and Federal, State and Local Government, March 9, 2006).

Community: Locations with high-quality indoor air are desirable places to live, play, and work and are ideal as community gathering spaces. Additionally, improvements to indoor air quality can reduce the incidence of diseases and ailments, directly improving the health of community members.

For schools and schoolchildren, good indoor environmental quality is even more urgent, due to the heightened sensitivity of young people to contaminants.

Strategies

Some strategies that project teams consider in addressing indoor air quality, along with examples of each, are as follows:

Strategies

Prohibit smoking.

The writing is on the wall—smoking causes lung disease, cancer, and heart disease as well as a host of other health problems. Secondhand smoke is no better. Project buildings can prohibit smoking within the building and restrict outdoor smoking to designated areas that won't cause smoke to irritate building occupants.

Ensure adequate ventilation.

We all know how draining it is to sit in a stuffy room. Well, it doesn't have to be stuffy and stale just because you're inside. Mechanical designers should ventilate with ample outdoor air to help ensure that indoor air contaminants are sufficiently diluted within the space.

Monitor carbon dioxide.

Monitoring carbon dioxide levels within spaces, especially spaces where the number of people changes, can allow for demand-control ventilation strategies in which the amount of air delivered to the space is controlled based on the needs of the users.

Install high-efficiency air filters.

High-efficiency air filters continually remove contaminants from the air and contribute to cleaner, healthier indoor air.

Specify low-emitting materials.

Low-emitting materials should be used rather than conventional products because of their reduced off-gassing of harmful contaminants.

Photo by Dale Photographic

Use integrated pest management.

Integrated pest management uses strategies that minimize or eliminate the potential for human exposure to pest-control chemicals by prioritizing nonchemical strategies, such as monitoring and baiting, rather than taking a "pesticide first" approach.

Strategies

Protect air quality during construction.

Project teams can implement an indoor air quality management plan during construction to improve the air quality for construction professionals as well as reduce the buildup of dust and other contaminants within the building's HVAC system. Additionally, projects that carefully protect absorptive materials from moisture damage reduce the potential for future mold growth.

Photo by Jennifer L. Owens

Conduct a flush-out.

Once construction is complete, before occupants move into the building, the space can be flushed with large quantities of outside air to remove residual contaminants from construction activities.

Employ a green cleaning program.

Green cleaning policies can greatly reduce the introduction of harmful chemicals during building operation and encourage best practices by custodial staff.

Thermal Comfort, Lighting, and Acoustics

Providing a high degree of thermal comfort and enabling occupants to play a role in controlling their environmental conditions lead to greater productivity and increased occupant satisfaction. Additionally, spaces that provide ample daylight, access to views, and high-performance acoustics further enhance the user experience.

Environment: Increased controllability fosters a partnership between the occupants and the building, and increases the likelihood that occupants will be advocates for operational practices such as turning off lights when they're not in use. Further, increased access to daylight can reduce the need for electric lighting during daytime hours, thereby decreasing energy use.

Economy: Personnel costs are a significant percentage of operating costs—much greater than energy or maintenance costs—thus, actions that affect employee retention, attendance, and productivity are significant for the bottom line.

Community: Most people prefer to live, work, and play in spaces where they have a degree of control over the indoor environment, are thermally comfortable, have access to daylight and views, and can communicate effectively due to good acoustics. Buildings that provide these comforts, especially if they openly accommodate public gatherings, are ideal as centers for the community they serve.

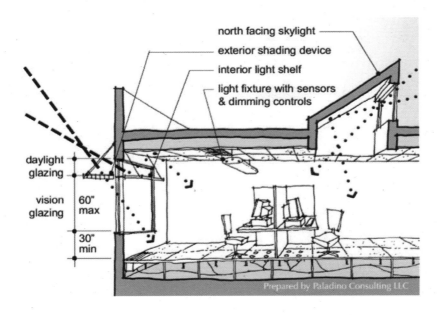

Strategies to increase occupant comfort indoors, along with examples of each, are provided below.

Strategies

Use daylighting.

Projects should be designed to introduce ample natural light into the space while providing glare-control devices to minimize the unwanted effects of unfettered sunlight. Areas that are not regularly occupied should be designed in the core of the building, while spaces such as offices and classrooms should be located along the building perimeter to take advantage of greater access to windows.

Photo by Eric Laignel

Install operable windows.

There is nothing quite like opening a window to get some fresh air. Operable windows provide occupants control over their environment so that on a mild day they can enjoy a breeze.

Photo by Anthony Semonti

Give occupants temperature and ventilation control.

Where operable windows are infeasible, occupant control over mechanically supplied heating, ventilation, and air conditioning should be provided.

Give occupants lighting control.

Task lighting at individual workstations and group lighting controls for shared spaces should be incorporated into the building to allow occupants to adjust the light levels to suit their needs and preferences. This also enables occupants to actively participate in energy savings by turning off lights when they're not needed.

Consider acoustics.

Effective verbal communication is integral to the way we live, work, and learn within buildings. Projects can make use of acoustical finishes, building geometry, duct insulation and other strategies to facilitate this critical component of human interaction.

Photo by Phil Weston

Conduct occupant surveys.

It's amazing how much you can learn just by asking a few questions. Asking occupants about their comfort within the space, and addressing thermal conditions, lighting, acoustics, and other elements contributing to their overall satisfaction, can help identify areas for improvement and foster occupant engagement in the operation of the building.

IEQ CATEGORY REVIEW

What are some key strategies for achieving high indoor environmental quality?

How do you balance ventilation and comfort versus energy efficiency? (For example, should we lower the thermostat in the winter months to save energy if that reduces comfort?)

Why is it important to provide lighting and thermal comfort controls to building occupants?

LEARNING ACTIVITIES

ASK AROUND

Ask your friends and co-workers about their working environments
– what do they like about them? What could make them better? Then consider
who would best be suited to address the improvements they have suggested - an
architect, interior designer, mechanical engineer, facility manager?

WALK AROUND

**Walk through an office or commercial building in your community to
identify features that improve indoor environmental quality.** What are
the best opportunities? List the current state of each feature as effective, neutral,
ineffective or absent.

FEATURE	EFFECTIVE	NEUTRAL	INEFFECTIVE	ABSENT
Daylighting				
Ventilation				
Operable windows				
Occupant control of lighting				
Occupant control of temperature				
Low-emissions materials				
High-efficiency air filters				
Green cleaning products and technologies				
Other				

INVESTIGATE

**Identify two building materials which are used within building
interiors** (e.g. indoor paint, finishes, furniture, etc.). Research the products to
identify all components and sub-components. Work in alone or with a group to
determine which components may emit harmful chemicals over their lifetimes.

PRACTICE QUESTIONS

1. **According to the Environmental Protection Agency, what percentage of time do Americans spend indoors?**

   a.) 75%
   b.) 65%
   c.) 90%
   d.) 50%

2. **Thermal comfort is typically attributed to what environmental factors?**

   a.) Temperature, humidity, and air speed
   b.) Ventilation, temperature, and daylight
   c.) Humidity, ventilation, and controllability
   d.) Density, temperature, and solar heat gain

3. **The abbreviation VOC refers to ___?**

   a.) Volatile organic compounds
   b.) Variable operating conditions
   c.) Variable ozone contaminants
   d.) Versatile organized composites

4. **Which strategy supports improved indoor air quality?**

   a.) Eliminate refrigerants with ozone-depleting potential.
   b.) Avoid the use of products with high carbon dioxide concentrations.
   c.) Use filters with a low minimum efficiency reporting value (MERV) rating.
   d.) Use materials with recycled content.
   e.) Use advanced framing techniques.
   f.) Design systems to deliver ample outside air.

5. **A school project in predesign would like to incorporate building strategies to maximize student learning. Which strategies should be considered to achieve this goal (select two)?**

   a.) Install individual thermal comfort controls in offices.
   b.) Incorporate daylight into classrooms.
   c.) Reduce the energy use of the building below the baseline standard.
   d.) Consider acoustical issues in core learning spaces.

6. **An operable window is considered what type of control?**

   a.) Lighting control
   b.) Thermal comfort control
   c.) Acoustical control
   d.) Environmental tobacco smoke control

7. **Demand-controlled ventilation is typically adjusted in response to ___?**

   a.) A time schedule
   b.) Occupant requests to the building operator
   c.) Carbon dioxide concentrations
   d.) VOC concentrations

8. **In addition to prohibiting smoking within the building, where is it also important to prohibit smoking to reduce occupant exposure to harmful airborne chemicals (select two)?**

   a.) In covered parking spaces
   b.) Near building entrances
   c.) Adjacent to building air intakes
   d.) In wooded areas

Answer Key on Page 137

KEY TERMS TO KNOW:

Cover up the left side of the page and test yourself to see if you can summarize the definitions of the following key terms. Or better yet, make flash cards.

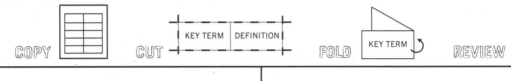

COPY   CUT   KEY TERM | DEFINITION   FOLD   KEY TERM   REVIEW

Key Term	Definition
Air Quality Standards	The level of pollutants prescribed by regulations that is not to be exceeded during a given time in a defined area. (EPA)
Ambient Temperature	The temperature of the surrounding air or other medium. (EPA)
ASHRAE	American Society of Heating, Refrigerating and Air-Conditioning Engineers.
Bake-Out	A process used to remove volatile organic compounds (VOCs) from a building by elevating the temperature in the fully furnished and ventilated building prior to human occupancy.
Carbon Dioxide Concentration	An indicator of ventilation effectiveness inside buildings. CO_2 concentrations greater than 530 parts per million (ppm) above outdoor conditions generally indicate inadequate ventilation. Absolute concentrations of greater than 800 to 1,000 ppm generally indicate poor air quality for breathing. CO_2 builds up in a space when there is not enough ventilation.

Commissioning (Cx)	The process of verifying and documenting that a building and all of its systems and assemblies are planned, designed, installed, tested, operated, and maintained to meet the owner's project requirements.
Commissioning Plan	A document that outlines the organization, schedule, allocation of resources, and documentation requirements of the commissioning process.
Commissioning Report	A document that details the commissioning process, including a commissioning program overview, identification of the commissioning team, and description of the commissioning process activities.
Contaminant	An unwanted airborne element that may reduce indoor air quality (ASHRAE Standard 62.1–2007).
Controllability Of Systems	The percentage of occupants who have direct control over temperature, airflow, and lighting in their spaces.
Daylighting	The controlled admission of natural light into a space, used to reduce or eliminate electric lighting.

Flush-Out	The operation of mechanical systems for a minimum of two weeks using 100 percent outside air at the end of construction and prior to building occupancy to ensure safe indoor air quality.
Indoor Air Quality	The nature of air inside the space that affects the health and well-being of building occupants. It is considered acceptable when there are no known contaminants at harmful concentrations and a substantial majority (80% or more) of the occupants do not express dissatisfaction. (ASHRAE Standard 62.1–2007)
Minimum Efficiency Reporting Value (MERV)	A rating that indicates the efficiency of air filters in the mechanical system. MERV ratings range from 1 (very low efficiency) to 16 (very high efficiency).
Off-Gassing	The emission of volatile organic compounds from synthetic and natural products.
Particulates	Solid particles or liquid droplets in the atmosphere. The chemical composition of particulates varies, depending on location and time of year. Sources include dust, emissions from industrial processes, combustion products from the burning of wood and coal, combustion products associated with motor vehicle or nonroad engine exhausts, and reactions to gases in the atmosphere. (EPA)
Pollutant	Any substance introduced into the environment that harms the usefulness of a resource or the health of humans, animals, or ecosystems. (EPA) Air pollutants include emissions of carbon dioxide (CO_2), sulfur dioxide (SO_2), nitrogen oxides (NO_x), mercury (Hg), small particulates ($PM2.5$), and large particulates ($PM10$).

Sick Building Syndrome (SBS)	A combination of symptoms, experienced by occupants of a building, that appear to be linked to time spent in the building but cannot be traced to a specific cause. Complaints may be localized in a particular room or zone or be spread throughout the building. (EPA)
Thermal Comfort	The temperature, humidity, and airflow ranges within which the majority of people are most comfortable, as determined by ASHRAE Standard 55–2004. Because people dress differently depending on the season, thermal comfort levels vary with the season. Control setpoints for HVAC systems should vary accordingly, to ensure that occupants are comfortable and energy is conserved.
Ventilation Rate	The amount of air circulated through a space, measured in air changes per hour (the quantity of infiltration air in cubic feet per minute divided by the volume of the room). Proper ventilation rates, as prescribed by ASHRAE Standard 62, ensure that enough air is supplied for the number of occupants to prevent accumulation of carbon dioxide and other pollutants in the space.
Volatile Organic Compounds (VOCs)	The amount of carbon compounds that participate in atmospheric photochemical reactions and vaporize (become a gas) at normal room temperatures, measured in grams per liter. VOCs off-gas from many materials, including adhesives, sealants, paints, carpets, and particleboard. Limiting VOC concentrations protects the health of both construction personnel and building occupants.

WHAT ABOUT INNOVATION IN DESIGN?

- What technologies have been developed in the past 10 years that are now commonplace but were once considered innovative?

- What is the goal of innovation?

- Can LEED recognize and encourage innovation?

- Can LEED reward the next generation of approaches to green building?

- Can a building be carbon neutral?

- Can a building achieve "zero net water use?"

- Can the design and operation of a building make people demonstrably healthier and more productive?

Read pages 83-84 of the Green Building and LEED Core Concepts Guide, Second Edition, or pages 65-67 in the First Edition.

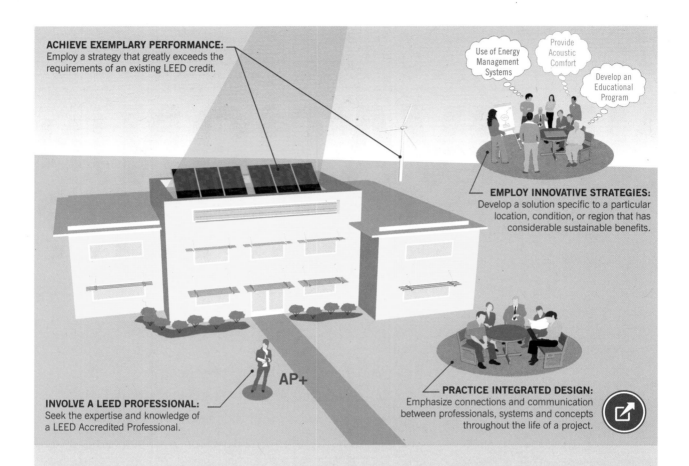

ACHIEVE EXEMPLARY PERFORMANCE:
Employ a strategy that greatly exceeds the
requirements of an existing LEED credit.

Use of Energy
Management
Systems

Provide
Acoustic
Comfort

Develop an
Educational
Program

EMPLOY INNOVATIVE STRATEGIES:
Develop a solution specific to a particular
location, condition, or region that has
considerable sustainable benefits.

AP+

INVOLVE A LEED PROFESSIONAL:
Seek the expertise and knowledge of
a LEED Accredited Professional.

PRACTICE INTEGRATED DESIGN:
Emphasize connections and communication
between professionals, systems and concepts
throughout the life of a project.

INNOVATION IN DESIGN OVERVIEW

After you've finished reading through all of the LEED categories and green
building strategies, hopefully you'll be excited and thinking about what's
next. Green building and sustainable design are all about evolving, learning, and improving. That's exactly why the Innovation in Design category
exists—to reward projects that go above and beyond what a credit requires or that incorporate a strategy that's not addressed by any prerequisite or credit.

There are two primary ways to earn an Innovation in Design credit:

● **Innovation in Design; and**

● **Exemplary performance.**

Lastly, a project team can earn one point for involving LEED Accredited
Professionals in the project.

Innovation in Design

As the building design and construction industry introduces new strategies for sustainable development, opportunities for additional environmental benefits will continue to emerge. Opportunities that are not currently addressed by LEED may include environmental solutions specific to a particular location, condition, or region.

With all sustainable strategies and measures, it is important to consider related environmental impacts. Project teams are encouraged to pursue opportunities that provide benefits of particular significance and should be prepared to demonstrate the environmental benefit of their innovative strategies.

Photo by Jenny Poole

Exemplary Performance

Project teams can earn exemplary performance points by implementing strategies that result in performance that greatly exceeds the level or scope required by an existing LEED prerequisite or credit. Exemplary performance opportunities are noted throughout the LEED reference guides. For example, if a project demonstrates a 45% reduction in potable water use, it is eligible for an Innovation in Design credit.

LEED Accredited Professionals

Multidisciplinary expertise in sustainable building principles and operational practices is recognized by the LEED Accredited Professional designation. LEED Accredited Professionals have the knowledge required to integrate sustainability into building operating and upgrade practices. The LEED Accredited Professional understands the importance of considering interactions between the LEED prerequisites and credits and their respective criteria. Building owners, facility managers, engineers, technicians, groundskeepers, purchasing staff, consultants, contractors, and others who have a strong interest in sustainable buildings are all appropriate candidates for accreditation.

ID CATEGORY REVIEW

How can involvement of a LEED Accredited Professional enhance a project?

What are some examples of today's innovative building methods that weren't in use five or 10 years ago?

What innovative strategies are on the horizon for five or 10 years from now?

LEARNING ACTIVITIES

THINK ABOUT IT

What strategies that might positively impact building performance aren't included in the current LEED Rating System?

If a project utilized a new technology that allowed them to drastically reduce the amount of materials that went into the construction of a building would it qualify for an Innovation Credit? What type of information would you need to know in order to document the benefits?

FIND OUR MORE

Visit the GBCI website (www.gbci.org) and research what is required to become a LEED Professional. Identify what the expectations are of the LEED Professional within an active LEED project. What role does the LEED Professional play?

INVESTIGATE

Investigate a couple of LEED projects in your area or in the USGBC case studies and identify which ID strategies were used.

PRACTICE QUESTIONS

1. **An Innovation in Design credit for exemplary performance would be available for ___?**

   a.) Implementing a comprehensive green building education program
   b.) Significantly diverting construction waste beyond the requirement of 75%
   c.) Achieving carbon neutrality
   d.) Providing 75% more spaces with daylighting
   e.) Developing a green housekeeping policy

2. **Projects can identify which credits have regional priority by ___?**

   a.) Checking the USGBC website
   b.) Researching locally pertinent environmental issues
   c.) Submitting a Credit Interpretation Request (CIR)
   d.) Calculating the net environmental impact

3. **An Innovation in Design credit for innovative strategies is likely available for projects that ___?**

   a.) Issue a press release to announce their LEED project registration.
   b.) Include two LEED Accredited Professionals on the project team.
   c.) Develop an educational outreach program.
   d.) Plan to enroll the project in LEED for Existing Buildings: Operations and Maintenance.

4. **A project team is considering pursuing an Innovation in Design credit for a strategy that will require a large capital expense. The team decides that it will implement the strategy only if the strategy will be eligible for an Innovation in Design credit. What steps should the team take to determine whether its strategy will be eligible for the credit?**

   a.) Research existing CIRs to see whether the strategy has been previously addressed.
   b.) Calculate the strategy's overall environmental impact and self-evaluate whether it is significant.
   c.) Draft a compelling narrative for the LEED submittal.
   d.) Check the LEED rating system.

Answer Key on Page 138

PRACTICE QUESTION ANSWER KEY

GREEN BUILDING ANSWERS

1. **b.** According to the U.S. Department of Energy's Energy Information Administration, 38% of all energy used within buildings is used for space heating. Lighting is the second largest user, at 20%.

2. **c.** In a conventional linear design, disciplinary boundaries can lead to inefficient solutions. Integrative design, conversely, brings together owners, operators, architects, engineers, and other stakeholders to identify cross-disciplinary opportunities to improve the environmental performance of building projects.

3. **a.** The conventional bottom line evaluates only economic prosperity. The triple bottom line, in contrast, is a way to evaluate performance relative to overall impact, including the human and environmental effects associated with performance.

4. **b.** Life-cycle assessment is a method used to analyze the environmental aspects and potential impacts associated with a product, process, or service. (LEED Reference Guide for Green Building Operations & Maintenance, 2009 Edition, page 499)

5. **a.** The predesign period of an environmentally responsive design process is the first phase in the development of a project. Bringing professionals together as early as possible in the process decreases the barriers to implementing innovative strategies such as passive thermal comfort and links between the site and the building.

6. **b.** LEED credits are weighted based on their relative importance. Credits that most directly address the most important environmental impacts and human benefits are given the greatest weight.

7. **c.** Life-cycle cost analysis considers the long-terms costs of building materials and design strategies.

8. **b & c.** Reducing the flow of water at hand-washing stations results in less water used per minute, thereby decreasing the total water used when washing hands. This in turn results in less hot water use, thereby decreasing the energy used to heat water.

U.S. GREEN BUILDING COUNCIL AND ITS PROGRAMS ANSWERS

1. **d.** GBCI recognizes three tiers of professional accreditation. Green Associate is the first level, followed by LEED AP with a specialty. Beyond LEED AP with a specialty, professionals can achieve the LEED Fellow designation.

2. **d.** USGBC is a nonprofit, member-driven body whose mission is "to transform the way buildings and communities are designed, built, and operated, enabling an environmentally and socially responsible, healthy, and prosperous environment that improves the quality of life." Its primary responsibilities are to develop and maintain technically robust building certification programs and educational offerings.

3. **a.** LEED for New Construction and Major Renovation addresses buildings, and

although furniture and appliances can be included within the scope of the program, it is not required to do so.

4. **a.** The carbon overlay is used to weight LEED credits based on the carbon footprint impact within a "typical" LEED building. The carbon overlay considers the impact of direct energy use, transportation, and the embodied emissions of water, solid waste, and materials.

5. **a.** Projects must first register with GBCI and pay the registration fee. Then, the LEED submittal documentation is compiled and submitted along with the certification fee. Once the review process is completed for both the design and construction credits, a final point tally (and achievement of all prerequisites) dictates the final LEED rating.

6. **d.** Once construction is complete, the project team can submit its construction phase credits (design phase credits may be submitted earlier). Only once the certification review is complete, and the project achieves the minimum number of points and all prerequisites, is the project certified.

7. **c.** The licensed-professional exemption allows registered professionals to forgo some of the documentation requirements.

8. **a, c, & d.** From the Guidelines for CIR Customers, the three correct answers should be tried before submitting a CIR. While LEED customer service representatives may be able to assist users in determining whether an issue is of appropriate technical complexity to be a CIR, they do not advise applicants as to whether a CIR is likely to be successful. Likewise, local chapters do not play a role in the CIR process. Finally, only one concise question or a set of related questions may be addressed within a single CIR, so it is not appropriate to include other issues within the same CIR.

SUSTAINABLE SITES ANSWERS

1. **c.** Vehicle miles traveled is a measure of transportation demand that estimates the travel miles associated with a project.

2. **c.** Decreasing impervious surfaces permits an increase in pervious surfaces. Pervious surfaces allow water to infiltrate the ground, thereby reducing stormwater runoff.

3. **d.** Nonpoint source pollution is caused by many sources, such as oil leaks in cars or fertilized landscape plantings, and is exacerbated by impervious hardscapes during heavy rainfall.

4. **a.** The solar reflectance index represents how well a surface rejects solar heat. Using materials with high solar reflectance indexes reduces the trapping of heat in site hardscapes, in turn reducing the heat island effect.

5. **a & d.** Increasing the floor-to-area ratio results in a smaller building footprint, thereby increasing the available site area to use as open space. Additionally, locating parking underground eliminates surface parking, likewise increasing the available site area that can be dedicated to open space.

6. **a & b.** Providing a carpooling incentive increases the percentage of occupants who co-commute, thereby reducing the total vehicle miles traveled. Upgrading the company fleet to hybrids or other low-emitting/fuel-efficient vehicles reduces the transportation impact per mile traveled, further lessening the environmental harm associated with automobile use.

7. **d.** Light trespass is the unwanted spillage of light onto adjacent properties. To minimize its impact on nocturnal environs, light trespass must be controlled.

8. **a.** The Sustainable Building Technical Manual: Part II recommends developing green design criteria before selecting the project site, in order to use the criteria to evaluate potential sites and guide the overall design effort.

WATER EFFICIENCY ANSWERS

1. **b.** Many water-using fixtures follow the Energy Policy Act (EPAct) guidelines for baseline water use. Other water-using fixtures follow standards such as the Universal Plumbing Code (UPC).

2. **a.** Projects that use reclaimed water for process uses reduce their demand on municipally supplied potable water.

3. **a & b.** Submeters and locally adapted plantings both contribute to water use reduction. Submeters ensure that water use can be tracked and leaks or overwatering quickly mitigated. The use of locally adapted plantings allows further reduction, because these plants are suited to the local climate and, once established, can be sustained with little or no ongoing irrigation.

4. **c.** Blackwater doesn't have a single definition that is accepted nationwide, although in all cases, water containing human waste is considered blackwater.

5. **d & e.** Nonpotable water is, by definition, not suitable for consumption, so although it is often acceptable to use for plant watering and waste transport, it is not usable for drinking, ice making, or bathing.

6. **c.** The Energy Policy Act (EPAct) establishes the baseline of 1.6 gallons per flush for all water closets.

7. **b.** Many municipalities supply both potable water (treated for human consumption) and nonpotable water (typically treated, but not to the same standard as potable water). This nonpotable water is often reclaimed from sources such as stormwater.

8. **a & d.** Process water is used for industrial systems such as HVAC, as well as for certain business operations, such as clothes washing and dish washing. Submetering can track consumption and allow for the early identification of inefficiencies or leaks. ENERGY STAR–certified clothes washers have low water factors, thereby ensuring that they use water efficiently.

ENERGY AND ATMOSPHERE ANSWERS

1. **b.** To reduce energy use, project teams should first reduce energy demand, followed by implementing energy efficiency measures such as installing high-performance equipment. Only then should renewable energy systems be considered. Finally, submetering equipment is used to maintain building performance.

2. **d.** Passive design strategies are used to reduce energy consumption by utilizing natural thermal processes (convection, absorption, radiation, and conduction) and sunlight to condition and light spaces.

3. **b.** RECs are tradable environmental commodities representing proof that a unit of electricity was generated from a renewable energy source. RECs are sold separately from the electricity itself.

4. **a.** The Lawrence Berkeley National Laboratory study found that commissioning for existing buildings has a median cost of $0.27 per square foot and an average simple payback of 0.7 years. For new construction, the study found that the median cost was $1 per square foot and had a payback of 4.8 years based on energy savings alone.

5. **a & c.** A refrigerant is weighted by its global warming potential and its ozone depletion potential.

6. **b & c.** Energy performance is a primary factor in determining how much energy is used per unit of cooling, thereby influencing the environmental impact of the HVAC system. Expected life is also a large driver of environmental impact, as systems that last longer need to be replaced less often, thereby decreasing the impacts associated with system replacement over the life of the building.

7. **c.** Process energy is defined as energy used to run office equipment, computers, elevators and escalators, kitchen cooking and refrigeration units, laundry washing and drying units, lighting that is exempt from the lighting power allowance, and miscellaneous items.

8. **b & c.** Renewable energy comes from sources that are not depleted by use. Examples include energy from the sun, wind, and small (low-impact) hydropower, plus geothermal energy and wave and tidal systems.

MATERIALS AND RESOURCES ANSWERS

1. **b.** Rapidly renewable resources, defined by LEED as having a planting/harvest cycle of 10 years or less, are replenished more quickly than conventional materials.

2. **c.** Waste reduction should be the first consideration of a waste management policy, followed by exploring opportunities for reuse. Recycling should then be considered for waste streams that can't be eliminated or repurposed.

3. **c.** The 50 tons sorted off-site are 60% diverted, resulting in 30 tons recycled. The 40 tons separated on-site are 100% diverted, resulting in 40 tons recycled. The

remaining 10 tons are incinerated, resulting in 0 tons recycled. The total recycled is 70 tons, which is 70% of 100 tons.

4. **c.** LEED has established 500 miles as the threshold for a regionally sourced material.

5. **b.** The EPA estimated greenhouse gas emissions from building waste and found that the United States currently recycles approximately 32% of its solid waste. Even a small increase in the overall recycling percentage could have enormous environmental benefits.

6. **b.** Postconsumer recycled content comes from consumer waste, contrasted with preconsumer recycled content, which comes from manufacturing processes.

7. **c.** Ongoing consumables are goods with a low cost per unit that are regularly used and replaced in the course of business (for example, paper, batteries, and soap).

8. **a, b, & d.** LEED addresses many environmental attributes of building materials, including their recycled content, regional harvest and manufacture, and reuse. LEED doesn't currently recognize materials based on their nation of origin or potential customizability.

INDOOR ENVIRONMENTAL QUALITY ANSWERS

1. **c.** The EPA estimates that Americans spend 90% of their time indoors, where concentrations of harmful contaminants may be dangerously high.

2. **a.** ASHRAE Standard 55 defines the environmental factors of thermal comfort as humidity, air speed, and temperature (air temperature and radiant temperature).

3. **a.** Volatile organic compounds (VOCs) are volatile at room temperature. Many of them are harmful to humans.

4. **f.** Delivering high volumes of outside air into the building interior dilutes indoor air contaminants, thereby improving indoor air quality.

5. **b & d.** Studies show that improved daylight in classrooms increases student learning, with one study showing 20% faster progression in math and 26% faster progression in reading. High-performance acoustics foster effective teacher–student and student–student communication.

6. **b.** Operable windows allow occupants to make adjustments to the air speed and temperature within the building, thereby controlling multiple environmental conditions for thermal comfort.

7. **c.** Demand-controlled ventilation modulates the delivery of outdoor air into the building based on occupancy. In many cases, outdoor airflow is increased when the building automation system detects increased carbon dioxide concentrations.

8. **b & c.** Smoking near building entrances or near building air intakes allows infiltration of environmental tobacco smoke into the building interior, thereby exposing occupants to harmful airborne chemicals.

INNOVATION IN DESIGN
ANSWERS

1. **b.** While many of these strategies may be eligible for an Innovation in Design credit, only diverting an increased percentage of construction waste is considered exemplary performance.

2. **a.** USGBC maintains a listing of all regional priority credits by ZIP code, which can be downloaded from the organization's website free of charge.

3. **c.** Innovative strategies expand the breadth of green building practices and introduce new ideas. They are distinct from exemplary performance innovation credits in that they address environmental issues not addressed elsewhere within the LEED rating system. Educational outreach programs are intended to educate building occupants about the building's features and the benefits of green building in general. These concepts aren't addressed elsewhere within the LEED rating system, so this strategy is a likely candidate for an Innovation in Design credit for innovative strategies.

4. **a.** Eligible innovative strategies aren't addressed within the LEED rating system or reference guides, but are instead evaluated via the CIR process. If the issue in question has never been addressed, a CIR is typically required to evaluate the strategy.

SAMPLE CHECKLISTS

LEED 2009 for New Construction and Major Renovation

Project Checklist

Project Name

Date

Sustainable Sites — Possible Points: 26

Y	N	?			
Y			Prereq 1	Construction Activity Pollution Prevention	
			Credit 1	Site Selection	1
			Credit 2	Development Density and Community Connectivity	5
			Credit 3	Brownfield Redevelopment	1
			Credit 4.1	Alternative Transportation—Public Transportation Access	6
			Credit 4.2	Alternative Transportation—Bicycle Storage and Changing Rooms	1
			Credit 4.3	Alternative Transportation—Low-Emitting and Fuel-Efficient Vehicles	3
			Credit 4.4	Alternative Transportation—Parking Capacity	2
			Credit 5.1	Site Development—Protect or Restore Habitat	1
			Credit 5.2	Site Development—Maximize Open Space	1
			Credit 6.1	Stormwater Design—Quantity Control	1
			Credit 6.2	Stormwater Design—Quality Control	1
			Credit 7.1	Heat Island Effect—Non-roof	1
			Credit 7.2	Heat Island Effect—Roof	1
			Credit 8	Light Pollution Reduction	1

Water Efficiency — Possible Points: 10

Y	N	?			
Y			Credit 1	Water Use Reduction—20% Reduction	
			Credit 1	Water Efficient Landscaping	2 to 4
			Credit 2	Innovative Wastewater Technologies	2
			Credit 3	Water Use Reduction	2 to 4

Energy and Atmosphere — Possible Points: 35

Y	N	?			
Y			Prereq 1	Fundamental Commissioning of Building Energy Systems	
Y			Prereq 2	Minimum Energy Performance	
Y			Prereq 3	Fundamental Refrigerant Management	
			Credit 1	Optimize Energy Performance	1 to 19
			Credit 2	On-Site Renewable Energy	1 to 7
			Credit 3	Enhanced Commissioning	2
			Credit 4	Enhanced Refrigerant Management	2
			Credit 5	Measurement and Verification	3
			Credit 6	Green Power	2

Materials and Resources — Possible Points: 14

Y	N	?			
Y			Prereq 1	Storage and Collection of Recyclables	
			Credit 1.1	Building Reuse—Maintain Existing Walls, Floors, and Roof	1 to 3
			Credit 1.2	Building Reuse—Maintain 50% of Interior Non-Structural Elements	1
			Credit 2	Construction Waste Management	1 to 2
			Credit 3	Materials Reuse	1 to 2

Materials and Resources, Continued

Y	N	?			
			Credit 4	Recycled Content	1 to 2
			Credit 5	Regional Materials	1 to 2
			Credit 6	Rapidly Renewable Materials	1
			Credit 7	Certified Wood	1

Indoor Environmental Quality — Possible Points: 15

Y	N	?			
Y			Prereq 1	Minimum Indoor Air Quality Performance	
Y			Prereq 2	Environmental Tobacco Smoke (ETS) Control	
			Credit 1	Outdoor Air Delivery Monitoring	1
			Credit 2	Increased Ventilation	1
			Credit 3.1	Construction IAQ Management Plan—During Construction	1
			Credit 3.2	Construction IAQ Management Plan—Before Occupancy	1
			Credit 4.1	Low-Emitting Materials—Adhesives and Sealants	1
			Credit 4.2	Low-Emitting Materials—Paints and Coatings	1
			Credit 4.3	Low-Emitting Materials—Flooring Systems	1
			Credit 4.4	Low-Emitting Materials—Composite Wood and Agrifiber Products	1
			Credit 5	Indoor Chemical and Pollutant Source Control	1
			Credit 6.1	Controllability of Systems—Lighting	1
			Credit 6.2	Controllability of Systems—Thermal Comfort	1
			Credit 7.1	Thermal Comfort—Design	1
			Credit 7.2	Thermal Comfort—Verification	1
			Credit 8.1	Daylight and Views—Daylight	1
			Credit 8.2	Daylight and Views—Views	1

Innovation and Design Process — Possible Points: 6

Y	N	?			
			Credit 1.1	Innovation in Design: Specific Title	1
			Credit 1.2	Innovation in Design: Specific Title	1
			Credit 1.3	Innovation in Design: Specific Title	1
			Credit 1.4	Innovation in Design: Specific Title	1
			Credit 1.5	Innovation in Design: Specific Title	1
			Credit 2	LEED Accredited Professional	1

Regional Priority Credits — Possible Points: 4

Y	N	?			
			Credit 1.1	Regional Priority: Specific Credit	1
			Credit 1.2	Regional Priority: Specific Credit	1
			Credit 1.3	Regional Priority: Specific Credit	1
			Credit 1.4	Regional Priority: Specific Credit	1

Total — Possible Points: 110

Certified 40 to 49 points   Silver 50 to 59 points   Gold 60 to 79 points   Platinum 80 to 110

MR Credit 4: Recycled Content

1–2 Points

Intent

To increase demand for building products that incorporate recycled content materials, thereby reducing impacts resulting from extraction and processing of virgin materials.

Requirements

Use materials with recycled content[1] such that the sum of postconsumer[2] recycled content plus 1/2 of the preconsumer[3] content constitutes at least 10% or 20%, based on cost, of the total value of the materials in the project. The minimum percentage materials recycled for each point threshold is as follows:

Recycled Content	Points
10%	1
20%	2

The recycled content value of a material assembly is determined by weight. The recycled fraction of the assembly is then multiplied by the cost of assembly to determine the recycled content value.

Mechanical, electrical and plumbing components and specialty items such as elevators cannot be included in this calculation. Include only materials permanently installed in the project. Furniture may be included if it is included consistently in MR Credit 3: Materials Reuse through MR Credit 7: Certified Wood.

Potential Technologies & Strategies

Establish a project goal for recycled content materials, and identify material suppliers that can achieve this goal. During construction, ensure that the specified recycled content materials are installed. Consider a range of environmental, economic and performance attributes when selecting products and materials.

---

1   Recycled content is defined in accordance with the International Organization of Standards document, ISO 14021 — Environmental labels and declarations — Self-declared environmental claims (Type II environmental labeling).

2   Postconsumer material is defined as waste material generated by households or by commercial, industrial and institutional facilities in their role as end-users of the product, which can no longer be used for its intended purpose.

3   Preconsumer material is defined as material diverted from the waste stream during the manufacturing process. Reutilization of materials (i.e., rework, regrind or scrap generated in a process and capable of being reclaimed within the same process that generated it) is excluded.

USGBC Policies

USGBC Logo Guidelines

According to the GBCI Green Associate Candidate Handbook, USGBC policies (trademark, and logo usage) are a category of exam content. Candidates should be familiar with the acceptable and unacceptable uses of USGBC Proprietary Marks.

Refer to www.usgbc.org/policies for the most up-to-date policies.